The Shadows of Creation

Michael Riordan, Science Information Officer at the Stanford Linear Accelerator Center and Staff Scientist at the Universities Research Association, received his Ph.D. in particle physics from MIT in 1973. He is the author of *The Hunting of the Quark* (1987), for which he won the 1988 American Institute of Physics Science Writing Award.

David N. Schramm, a renowned theoretical astrophysicist, is Professor of Physics at the University of Chicago. He received his Ph.D. from the California Institute of Technology in 1973. Schramm has been a guiding force behind much of the recent development in theoretical astrophysics, and has written many books and articles on the subject, including *From Quarks to the Cosmos* (1989), which he co-authored with Leon Lederman.

The Shadows of Creation

Dark Matter and the Structure of the Universe

MICHAEL RIORDAN
and
DAVID N. SCHRAMM

OXFORD UNIVERSITY PRESS

In memory of
Andrei Sakharov and Yacov Zeldovich

Oxford University Press, Walton Street, Oxford OX2 6DP
Oxford New York Toronto
Delhi Bombay Calcutta Madras Karachi
Kuala Lumpur Singapore Hong Kong Tokyo
Nairobi Dar es Salaam Cape Town
Melbourne Auckland Madrid
and associated companies in
Berlin Ibadan

Oxford is a trade mark of Oxford University Press

First published 1991 by W. H. Freeman and Company
First issued as an Oxford University Press paperback 1993
Reprinted 1993

British Library Cataloguing in Publication Data
Data available
ISBN 0-19-286159-X

Printed in Great Britain by
Biddles Ltd
Guildford and King's Lynn

Table of Contents

––––––– •●• –––––––

Foreword

I was in high school when I first heard that the light from distant galaxies was shifted towards the red end of the spectrum. I was told that this meant that the galaxies were moving away from us and that the Universe was expanding, but I didn't believe it. I thought that there must be some other explanation for the redshift of the light. Maybe light got tired and redder traveling that great distance. An essentially static universe seemed much more natural than an expanding one. Where would it be expanding to, and what could have set it expanding?

There is in fact a good reason why the Universe should not be static. The law of universal gravitation, proposed by Newton in 1687, states that every body in the Universe attracts every other body. As a result, if these bodies were at rest relative to each other at one time, they would start to fall toward each other. That might make it sound like gravity would cause everything in the Universe to get closer together, rather than moving farther apart. However, one could equally well suppose that objects in the Universe were moving away from each other, but that the rate of separation was being slowed down by gravity. The important point is that the Universe has to be either expanding or contracting. It cannot remain static if gravity is attractive.

According to our understanding of the early Universe, objects should be moving apart at a rate which is just fast enough that gravity will never quite make them stop and start falling together again. But they shouldn't be moving away from each other any faster than this critical rate, which will depend on the amount of matter in the Universe.

We can measure the rate at which objects are moving apart, but the amount of matter we can see in the Universe is not enough to slow down this rate of expansion significantly. So there are two possibilities; either our understanding of the very early Universe is completely wrong, or there is some other form of matter in the Universe that we have failed to detect. The second possibility seems more likely, but the required amount of missing "dark" matter is enormous; it is about a hundred times the matter we can directly observe. This book is about the missing matter and the form it might take. It is about the search for the major part of the Universe; the part we actually see can be only a tip of the iceberg, to mix a metaphor. What can this invisible constituent be? Is it black holes or some new and unknown kind of elementary particle? The quest is on for finding 99 percent of the Universe.

Stephen Hawking
May 2, 1990

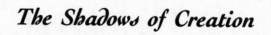

The Shadows of Creation

1

The Big-Bang Universe

One evening late in 1979, Alan Guth made a discovery that was to change the course of modern physics. A 32-year old scientist at the Stanford Linear Accelerator Center in Menlo Park, California, he had recently become intrigued by the connections between elementary particles — the tiniest things at the very base of existence — and cosmology, the study of the largest conceivable structures. What he learned from his equations that dark December night set a chain of events in motion that would ultimately revamp our ideas about the entire Universe.

Guth was trying to solve a nagging problem then bothering an obscure group of scientists who were attempting to marry these two disciplines. If their ideas were correct, extremely heavy particles (called "magnetic monopoles") should have existed in great profusion. The Universe should have been *swarming* with them, in fact, but not a single one had ever been observed. Not one.

What Guth accomplished that fateful night was to find a natural way to make the unwanted particles vanish, thereby resolving the troubling discrepancy. If immediately after its birth the Universe had exploded, expanding convulsively in every possible direction, then space itself would have swelled to such an enormous extent that these ponderous corpuscles would be essentially nonexistent. Only a stray few might remain in all the trillions upon trillions of cubic light-years visible from Earth. No wonder nobody had ever found any!

That, however, was only the beginning. Guth's primordial explosion, which he dubbed "inflation" after the persistent monetary problem of the 1970s, made a lot of other things fall into place. Inflation found a receptive audience among the physicists working on cosmology; it became gospel almost overnight.

Inflation also had a startling, inescapable consequence. If it were true, then there had to be far more matter in the Universe than anybody had previously imagined. Almost a *hundred* times more than astronomers could see with their telescopes!

During the 1970s, there had been growing indications that something was amiss in the heavens. More matter appeared to be lurking in the depths of space than met the eye — stuff that somehow didn't shine but still had an impact upon how stars moved and galaxies whirled. "Dark matter," as it came to be called, seemed to be an unseen voyager of the great, yawning voids in space.

Until Guth's landmark discovery, however, few scientists had ever imagined there could be so *much* dark matter. Studies of the heavenly abundances of the light elements — hydrogen, helium, and lithium — gave one clear limit. If the dark matter were made of normal atoms like those we encounter on the face of the Earth, then there might be 10 or 15 times as much of it in the Universe as the visible part, perhaps, but no more. One hundred times more was out of the question.

Inflation, then, meant that the majority of matter in the Universe — 90 percent of it or more — is absolutely and totally different from the everyday stuff that we walk on and eat with and breathe. This exotic dark matter doesn't emit or absorb or even reflect light. It cannot be made of the atoms and molecules most of us take for granted. And it may be the commonest substance in the Universe. *We* are the ones made of unusual stuff.

This dramatic realization has emerged from a virtual revolution in cosmology that has taken place since the mid-1960s. Many observations and experiments lend overwhelming support to a story of creation — the Big Bang — with which most scientists now concur. They may differ on the exact details, but almost all would agree that our Universe began in a tremendous eruption some 10 to 20 billion years ago. At that moment all its matter and energy were born; even space and time came into being.

Perhaps the greatest mystery remaining in this Big-Bang picture of the Universe is the dark-matter question. Previously thought to be an empty void, except for a sprinkling of visible stars, gas, and dust, space now seems to be filled with an almost ineffable mist that pervades much of its vast extent.

What is this dark matter? How did it come into being? And how does it affect the visible objects in the Universe? These are the questions that intrigue physicists and cosmologists today.

———————————— • ● • ————————————

Cosmologists of previous eras have repeatedly had difficulty with a void. Something within the human heart recoils at the thought of absolute nothingness. In their unceasing efforts to discern a rational order amidst the apparent chaos of the heavens, people have appealed to unseen beings or substances lurking behind the visible veneer of existence. Something *else* had to be out there in space, propping up the luminous orbs and determining their motions and the patterns they trace.

To primitive cultures, the night sky harbored invisible beings and influences thought to govern earthly events. All kinds of gods and spirits supposedly inhabited this blackness—revealing their presence only through the sundry constellations that stretched across ancient skies.

The more sophisticated Greeks pictured their heavens as a realm of unearthly perfection, filled with an ideal substance called the "aether." Very different stuff from their four basic elements—earth, air, water, and fire—the aether was to them an eternal, unchanging fluid, a "fifth substance" in which the Sun, Moon, planets, and stars made their orderly processions. Aristotle crystallized this aether, making it a solid rather than a fluid. He taught that all celestial bodies revolved about the Earth on stately spheres of hard, transparent crystal.

Working in Alexandria during the second century after Christ, the astronomer Claudius Ptolemy took Aristotle's crystal spheres one step further. To explain the slightly erratic, wandering motions of the planets, he added smaller spheres that rolled upon the larger ones surrounding the Earth. The planets moved upon these smaller spheres in what were called "epicycles" and "eccentrics." Ptolemy's model of the Universe actually fit the observed motions of the planets amazingly well—so well, in fact, that his ideas held sway in the Mediterranean world for more than a thousand years.

In 1543, the Polish astronomer Nicolaus Copernicus finally published his masterwork, *De Revolutionibus*, in which he described a heliocentric model of the Universe—one in which the planets revolved about the Sun, not the Earth. His act of removing humanity from the center of the Universe, viewed as heretical in almost all of Christian Europe, had tremendous theological and philosophical implications. The book became the focus of controversy between secular and ecclesiastical authorities for more than a century.

With funding that would make a modern scientist envious, the Danish nobleman astronomer Tycho Brahe assembled the most accurate sighting instruments in existence during the late 1500s, before the invention of the telescope. The vast quantity of precise data he collected on the motions of the planets was eventually used by his assistant, Johannes Kepler, to develop a variation of the Copernican model of the Universe. In this version the planets moved along elliptical paths rather than in the ideal, perfect circles held dear for centuries by the ancient Greeks and their intellectual heirs. Kepler also used Brahe's data to develop two other empirical laws of planetary motion.

The deeper, underlying physical explanation for Kepler's empirical laws was discovered by a succeeding generation of seventeenth-century scientists and philosophers. In that era, now called the Age of Reason, the scientific method was born. In 1687, Isaac Newton, a reclusive, mystical English physicist and mathematician ensconced at Cambridge University, announced his principle of universal gravitation based on Kepler's laws. The force of *gravity* was what caused the planets to trace out their elliptical orbits about the Sun.

Newton and other leading scientists of his day effectively traded the crystal spheres of Aristotle and Ptolemy for an empty void. The space between the celestial bodies became completely vacant. What suspended these orbs and governed their motions was not a substance like the aether but a universal force. Newton's law of gravitation provided a complete framework in which to understand not only the motion of planets around the Sun but also moons around planets, apples falling from trees, and, eventually, galaxies orbiting each other—all within empty space.

In the nineteenth century the aether made a brief comeback as the all-pervasive medium through which light waves supposedly raced as speedy vibrations. The "luminiferous aether," as it was called, was fantastic stuff—lighter than air and stiffer than steel. It arose in a theory developed by the Scottish physicist James Clerk Maxwell, who combined electricity and magnetism into a

single, unified, "electro-magnetic" force. One consequence of
Maxwell's famous equations was the existence of electromagnetic waves, of which light is one form. But what were these
waves moving in? Because a wave is a disturbance, like a ripple
upon the surface of a pond, there had to be some kind of
medium (such as the water in the pond) that oscillated as light
waves passed through it, or so most physicists imagined during
the late 1800s. Thus was born the luminiferous aether — a ubi-
quitious medium that supposedly transmitted light.

But repeated attempts to detect this aether directly ended in
failure. In 1905, Albert Einstein finally showed it was completely unnecessary. His theory of special relativity allowed
light to travel effortlessly through empty space forever; this
theory also banished absolute length and time from the Universe. Space became emptiness again, filled only with shining
lumps called stars (and great swarms of them called galaxies)
that were very sparse indeed.

During the past 20 years, however, scientists have again
begun to suspect that there *is* much more "out there" than the
sparse concentrations of luminous objects visible in telescopes.
No longer is space viewed as an empty, formless void. It seems
instead to harbor a kind of invisible, almost ineffable substance
that actually contributes the majority of material in the Universe. So strange is this dark matter, however, that it may be
presumptuous to call it "matter" at all. It is the aether of today.

——————————— • ● • ———————————

Until the 1920s, most astronomers and cosmologists thought the
Universe was a static entity consisting only of the stars that
constitute our own galaxy, the Milky Way. This neat picture was
abruptly shattered by the pivotal work of Edwin Hubble. A
championship basketball player and Rhodes scholar who aban-
doned the practice of law to pursue astronomy at the University
of Chicago, Hubble proved that wispy celestial objects then

called "nebulae" were in fact *external* to the Milky Way. They are separate galaxies themselves, vast clumps and whirlpools containing many millions of stars. Soon thereafter, he observed that these galaxies were often grouped into "clusters" — enormous archipelagos, each including many galaxies apiece.

While examining the brightest galaxies in these clusters, Hubble noticed something odd about the light they emitted. Depending on its surface temperature, a star emits a characteristic spectrum of electromagnetic radiation that peaks in the visible portion of the spectrum. The Sun, for example, appears slightly yellowish because its spectrum reaches its peak at wavelengths

Albert Einstein and Edwin Hubble in 1930. Einstein is peering through the 100-inch telescope on Mt. Wilson, which Hubble used to discover that galaxies are receding from the Milky Way.

corresponding to yellow light. But the light from a distant galaxy, Hubble observed, had generally been shifted into the redder part of the spectrum, to longer wavelengths. In other words, the distance between successive wave crests had grown — as if the light waves from these galaxies had somehow been stretched out, like a spring pulled apart, during their long trip to Earth.

The amount of change in the wavelength of light from a galaxy, called its "redshift," gave Hubble a way to measure the galaxy's velocity. (The highway patrol uses a similar principle to catch speeders, using radar.) A stationary atom will emit or absorb light at a group of characteristic wavelengths called its "atomic lines." Sodium, for example, has two prominent yellow lines in its emission spectrum. Hubble and his associates measured the wavelengths of atomic lines in the light of distant galaxies and found that they were almost always *longer* than those of stationary atoms, as observed in the laboratory.

Such redshifts implied a high-speed motion of the light source *away* from our own galaxy. (In the same manner, the sound of the whistle on a train moving away from us is shifted to a lower pitch, or longer wavelength.) Hubble's detailed studies indicated that the redshift grew larger as the distance to a galaxy increased, a fact he published in a famous 1929 paper. The deeper he peered into space, the faster the distant galaxies seemed to be fleeing the Milky Way. The Universe was expanding.

The most difficult part of Hubble's analysis was determining the distance to another galaxy or to a cluster accurately. On the surface of the Earth, we determine distances using meter sticks and triangulation. We can measure the distance to the Sun or the planets accurately by determining the time it takes for radar pulses to bounce off them and return to Earth — and then multiplying this time interval by the speed of light. But even the nearby stars are too far away to use radar, and the distances to only the closest ones can be determined by triangulation, using the apparent shift in their position as the Earth orbits the Sun. When it comes to galaxies, these techniques are hopeless.

What Hubble did to overcome these limitations was to build up a distance scale using the intrinsic brightness of certain very bright stars called "Cepheid variables," which could be seen in nearby galaxies like the Andromeda nebula (also known as M31). By comparing the apparent brightness of these stars with their intrinsic value, he established the distance to the nearby galaxies. (A light source appears dimmer the farther away it is. Its apparent brightness decreases as the square of this distance.) This was the technique, in fact, that Hubble used to prove that spiral nebulae such as M31 were too far away to be part of the Milky Way and had to be separate galaxies.

This technique could only be used for nearby galaxies, in which the Cepheid variables can be resolved as individual stars and therefore be used as reference points, or "standard candles." Beyond several million light-years (a light-year is the distance light travels in a year, about 6 trillion miles or 9 trillion kilometers) this technique became impossible, and Hubble had to resort to other methods. To determine the distance to faraway clusters, he used the brightest galaxies in them as a standard candle; by comparing their apparent brightness with a standard value, he could estimate these distances.

In his 1929 paper, Hubble showed that the recessional velocity v of a cluster is proportional to its distance d from the Milky Way, or $v = Hd$. The constant of proportionality H is now known as the "Hubble constant." It gives the *rate* at which the Universe is expanding, and this relationship indicates that the expansion is uniform in all directions.

The Hubble constant applies to all clusters of galaxies during the present epoch of the Universe. When we look at extremely large distances, however, the light hitting our telescopes has taken so long to reach us that we are actually observing how these clusters looked many millions and even billions of years ago. Thus, the Hubble "constant" determined for that epoch is different from the value it has in the present one (although such a difference is difficult to establish, in practice). It measures the rate of expansion in that previous epoch—a rate that was

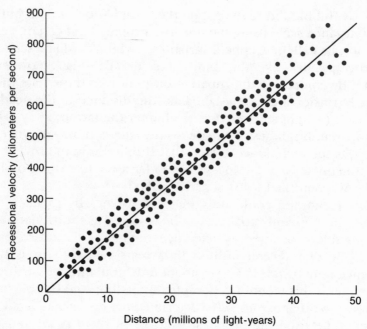

Linear relationship between recessional velocity and distance to a galaxy, which was discovered by Edwin Hubble in 1929. The slope of the line is known as the Hubble constant *H*.

actually *faster* millions of years ago than it is today. The expansion of the Universe is slowing down.

• ● •

The next major advance in cosmology came in the 1960s, when the cosmic background radiation was first observed. This soft, all-pervasive glow of microwave radiation was discovered accidentally in 1964 by Arno Penzias and Robert Wilson, who were then working at the Bell Telephone Laboratory in Holmdel, New Jersey. These two young radio astronomers had been given ac-

Robert Wilson (left) and Arno Penzias, who discovered the cosmic background radiation in 1964. Behind them stands the microwave antenna they used in their epochal work, which altered the course of modern cosmology.

cess to a large, horn-shaped antenna originally developed for tracking the early Echo and Telstar communication satellites. While trying to adapt this device for radio astronomy, they noticed a steady hum of radio noise in their equipment.

No matter which way Penzias and Wilson pointed the horn, the hum was still there. And its intensity was well above the designated noise level of the antenna. Thinking it only a bothersome "background" at first, and not a true signal, they spent months trying to eliminate it, without success. They even climbed into the horn and scrubbed away some smelly pigeon droppings that had accumulated. Still the hum persisted. This "noise" behaved exactly like the signal that would have been observed if microwave radiation were coming uniformly from all directions in space. It corresponded to what would have been observed from a black body cooled to a temperature of three degrees above absolute zero, or 3 degrees Kelvin (3°K).

Unknown to Penzias and Wilson, such a soft, uniform glow had already been predicted in 1948 by the whimsical Russian-American physicist George Gamow and his colleagues Ralph Alpher and Robert Herman. It should be the remnant, they suggested, of the ultrahot, intense radiation that would have been released if the Universe had begun in a stupendous explosion. As the Universe expanded, this radiation would have cooled down to several degrees Kelvin, which corresponds to the microwave portion of the electromagnetic spectrum.

At about the same time as Penzias and Wilson were struggling to get rid of the "noise" in their equipment, Princeton physicists Robert Dicke, James Peebles, and David Wilkinson began building their own radio antenna. Their aim was to search for the cosmic background radiation predicted by Gamow and colleagues. After learning of the Bell scientists' serendipitous discovery, they quickly finished their antenna and confirmed this epochal find. There was indeed a soft glow of microwave radiation, corresponding to a temperature of about 3°K, coming from all directions.

Until the early 1960s, there had been two principal cosmological theories that attempted to explain the expansion of the Universe observed by Hubble in the 1920s. The "steady-state" theory, authored by Cambridge University astrophysicists Fred Hoyle, Hermann Bondi, and Thomas Gold in the late 1940s, held that matter was being generated continuously to fill up the yawning voids that were gradually emerging as space expanded. The Big-Bang idea espoused by Gamow and collaborators (the name for which was supplied by Hoyle in a moment of mirthfulness on a BBC radio program) contended that all matter and energy in the present Universe had been created at a single instant billions of years ago, in an immense, primeval fireball.

Before Penzias and Wilson's discovery, these two rival cosmological theories had been on an equal footing. There were no convincing observational tests to help scientists decide in favor of one theory or the other. But the cosmic background radiation was difficult to account for in the framework of a steady-state universe, while it was a natural outcome of the Big Bang.

To understand the origin of this radiation better, note that any continuous body of matter — such as your own body, the wall of a room, or the Sun — can be described as having a characteristic temperature. This temperature is $98.6°F$ for your body, about $70°F$ for the wall, and about $5000°K$ for the surface of the Sun. When the Universe was squeezed together so that it was a dense plasma, as happened during the Big Bang, it too had its own characteristic temperature. Every individual object today has a different temperature, of course, but back in the distant past the Universe had a single temperature at any given moment.

An object with a characteristic temperature gives off a particular spectrum of electromagnetic radiation that depends only on that temperature. The hotter the body, the higher the frequency and the shorter the wavelength of the electromagnetic waves emitted. Heat an iron poker in the fireplace, and it will begin to glow red, then yellow, and finally white as it becomes hotter and hotter.

Because any radiation emitted by the dense, hot plasma of the Big Bang could never escape the Universe, it had to be still travelling through space. Although the apparent temperature of the radiation could have cooled, because the Hubble expansion would have stretched short wavelengths into long, it must nevertheless continue to exist. Because energy must be conserved, the radiation could not simply disappear. This is the essence of the arguments made by Gamow and colleagues, but it was almost 20 years before Penzias and Wilson found evidence for the leftover radiation—a dim relic of the most colossal explosion ever.

Further evidence for the Big Bang began to accumulate in the mid-1960s. At about the time of Penzias and Wilson's discovery, Hoyle, Peebles, and other scientists proved that such a hot primeval blast would have cooked about a quarter of the mass in the known Universe into helium, the lightest element after hydrogen. The natural abundance of this element observed by astronomers came in very close to the prediction, lending added support to the Big-Bang idea.

In steady-state cosmology, the Universe was never any denser than it is today, so the only way heavier elements could be built up from hydrogen is by thermonuclear burning in the cores of stars. In the Sun, for example, atoms of hydrogen are continually fusing to create atoms of helium, giving off energy in the process. Eventually the helium itself fuses to make carbon and oxygen. Such thermonuclear burning in massive stars produces the bulk of heavy elements, from carbon and oxygen through iron. But the sum of all such elements heavier than helium makes up only about 2 percent of the visible mass of the Universe. Thus, it is extremely improbable that thermonuclear burning in stars could ever have produced so much of a single element like helium (which contributes almost 25 percent of the observable mass of the Universe).

The remaining advocates of the steady-state cosmology had one last argument against the Big Bang. They noted, quite rightly, that the spectrum of cosmic background radiation had to have a specific shape corresponding to the radiation emitted by a blackbody, an ideal body that absorbs any radiation that hits it and reflects none. Such a blackbody spectrum, which had been predicted by Gamow, would have a peak at a wavelength of about 1 millimeter. In the 1960s, the only reliable measurements had been made at longer wavelengths, where the radiation could penetrate the atmosphere and reach ground-based antennas.

By 1974, however, the cosmic background radiation had been sampled at many different wavelengths. Particularly important in this effort was the work of Paul Richards at the University of California, Berkeley. Using balloon-borne radio antennas that soared well into the stratosphere, he was able to study microwave radiation at wavelengths that are absorbed by the Earth's atmosphere. Others placed their antennas on high-flying jets and suborbital rockets. These high-altitude observations finally confirmed that the detailed spectrum of the background radiation

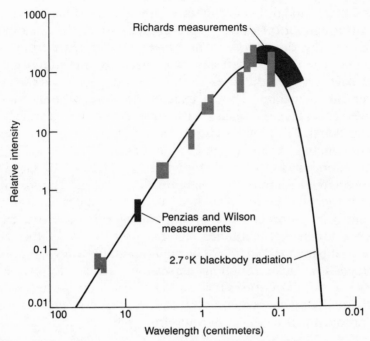

Spectrum of cosmic background radiation, as measured by the mid-1970s. The spectrum peaks at a wavelength of about 1 millimeter (0.1 cm) and has a shape like the spectrum emitted by a blackbody at 2.7°K.

indeed peaked near 1 millimeter and had the required blackbody shape. Big-Bang cosmology was here to stay. Steady-state ideas fell into oblivion.

• ● •

The steady "Hubble expansion" of the Universe has had tremendous implications for cosmology. To understand the phenomenon better, imagine the Universe as if it were a huge mass of raisin-bread dough baking in an enormous oven. Raisins sprinkled throughout the dough represent individual clusters of galaxies. As the dough rises, the raisins spread apart from one another. The farther apart any two raisins are, the more dough they have between them, expanding with the heat, and thus the faster they separate. This is exactly what we witness in the Universe; replace the raisins with clusters of galaxies and think of the dough as the very fabric of space itself, which has been stretching and expanding since time began.

An important aspect of this kind of expansion is the fact that shapes and configurations remain the *same* as time elapses; they merely become larger. The loaf still looks like a loaf as the dough rises. In other kinds of expansions, this is not the case. After a firecracker explodes, for example, it looks like tiny shreds of paper and shards of burned-out gunpowder — not like a bigger firecracker. Such an expansion does not follow Hubble's law, which requires that the relative velocity between two individual objects, whether raisins or clusters, must always be proportional to their distance from one another. The Hub-

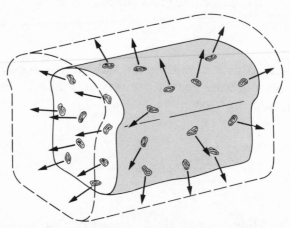

A "raisin-bread" model of the Universe. As the loaf rises, it maintains the same general configuration with the raisins moving steadily apart from one another. The farther apart two raisins are, the faster they separate.

ble expansion is a very smooth, steady expansion, free of turbulence.

Now suppose we wish to look *backward* in time and see how the Universe appeared at earlier moments. Actually you do this any time you gaze into outer space. The light arriving today from even a nearby galaxy like M31 took millions of years to reach Earth. This light brings information about what happened in that galaxy during our prehistoric era, when this ancient light was emitted. With a powerful telescope, we can stare *billions* of years backward in time, picking up dim remnants of light that departed long ago from the most distant objects visible today.

In a Big-Bang Universe, however, there is a limit, called our "horizon," to how far we can see. We can see only as far as light could have traveled in the 15 billion years or so that have elapsed since the primeval fireball cooled. (Time as we know it actually *began* with the Big Bang.) If we draw an imaginary sphere about our galaxy, with a radius equal to this distance — about 15 billion light-years — the surface of the sphere represents our horizon. An observer in the Milky Way can perceive only events happening *within* that sphere, not beyond it.

Any observer in the Universe has a similar horizon. To understand this, think of each and every raisin in our ball of dough as being surrounded by an imaginary sphere, all with the same radius. As time elapses and the dough expands, these horizon spheres grow too, each enveloping more raisins and more dough. This concept of a horizon is unique to Big-Bang cosmology. (Horizons would have no meaning in a steady-state cosmology because it possesses no origin of time.)

In the early Universe, the horizon distance was far shorter than it is today. When it was only 1 year old, for example, light could have traveled only a single light-year, so the horizon then was 1 light-year. Any two objects closer than that distance would have been in contact with one another. All objects farther than 1 light-year apart could have had no knowledge of each other's existence. But as time elapsed and these horizons grew, they would eventually come within eyeshot of one another.

Now imagine further that we can (by some unfathomable means) extract ourselves from the Universe, get to a viewpoint where we can observe the happenings of everything all at once. Of course this is absurd. It corresponds to our escaping altogether from space and time. But it is a helpful imaginary perspective that we have actually been using here when we speak of the Universe as a rising loaf of raisin bread with clusters of galaxies sprinkled throughout.

From this perspective it is natural to ask what happens if we reverse time and let the Universe contract. What happens if the loaf suddenly begins to fall instead of rise? The clusters of galaxies (raisins) get closer and closer together until they begin touching and overlapping. Everything gets hotter and hotter, because the same total amount of energy is being crammed into a smaller and smaller volume. The density of matter grows steadily as the volume decreases.

Eventually things get so hot and so dense that matter breaks up into a seething broth of subatomic particles, what physicists call a "plasma," in which nuclear and elementary-particle physics take over. This is exactly what happened in the Universe during the first few minutes of its existence. At the earliest split second, we suspect, all space and time, all energy and matter emerged from a perfectly symmetric condition we can experience today only in abstract mathematical equations.

The history of the Universe from this earliest instant has been a saga of ever-growing asymmetry and increasing complexity. As space expanded and temperatures cooled, particles began collecting and structures started forming from this ultrahot, featureless plasma. Eventually clusters and galaxies, stars, planets, and even life itself emerged. In at least one corner where conditions were ideal, intelligent beings evolved to the point where they could begin to comprehend these fantastic origins. Above all else, the Big-Bang Universe is an *evolutionary* Universe fraught with possibilities.

———————————— • ● • ————————————

Despite its orderly Hubble expansion, the Universe has gone through some remarkable changes since it began about 15 billion years ago. When it was 1 second old, it was filled uniformly with subatomic particles — the electrons, protons, and neutrons that make up ordinary matter, plus particles of light called "photons." There were also large numbers of "positrons," the antimatter opposites of electrons, and vast hordes of extremely light particles called "neutrinos" of which we will say much more in the next chapter.

The temperature of the Universe 1 second into its evolution was about 10 billion degrees (or 10^{10} °K, in scientific notation), which is the temperature at which atomic nuclei disintegrate. This is about a thousand times hotter than the core of the Sun, and a million times hotter than its surface.

The density of the Universe — the total amount of matter and energy packed into a given unit volume — at that moment was about 10 kilograms per cubic centimeter. For comparison, water has a density of 1 gram (a thousandth of a kilogram), and lead about 10 grams, per cubic centimeter. Thus, the Universe at an age of 1 second was about a thousand times denser than lead. A chunk of it about the size of a tennis ball would have weighed as much as an automobile!

By the time the Universe was a few minutes old, all the neutrons had combined with protons to make heavier nuclei like helium, and the positrons had annihilated most of the electrons to generate still more photons. A sprinkling of protons and electrons still permeated this fiery plasma. The temperature had fallen to several hundred million degrees, still much hotter than the core of the Sun, and the density had dropped to about a hundredth of a gram per cubic centimeter (heavier than air but lighter than Styrofoam). A large shopping bag full would have weighed less than a pound.

Not much in the way of change occurred for the next 100 thousand years or so. The Universe continued its steady Hubble expansion, gradually cooling off until its temperature had reached a few thousand degrees (this is about the temperature of melting brimstone — or sulfur — which medieval Christian theologians thought to be the temperature of Hell). At that point its density was only about a billionth of a trillionth of a gram per cubic centimeter (or 10^{-21} g/cm³), which is far less dense than the best vacuum attainable in terrestrial laboratories. An earth-sized balloon full of this stuff would weigh around 2 tons, or about as much as a small truck.

At this time the electrons permeating the Universe finally slowed down to the point where they could be captured by the much more ponderous nuclei, forming a gas of electrically neutral atoms of hydrogen and helium — plus a small trace of lithium, the next lightest element. Prior to this moment, the Universe had been an incandescent plasma like that which exists inside neon lighting fixtures. Photons had been trapped within this plasma because they interacted readily with all the charged particles in it. When these charged particles became bound up in electrically neutral atoms, however, photons suddenly could

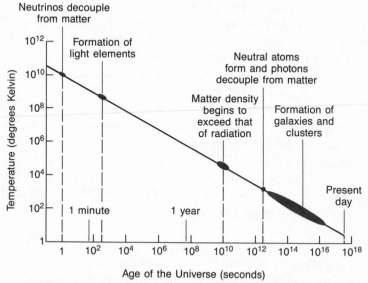

Time evolution of the average temperature of the Universe, from the moment it was 1 second old to the present day. As the Universe expands, both its temperature and density fall steadily.

break free and roam off to infinity in all directions, like light passing through thin air. Eons later, Penzias and Wilson would finally detect the dim remnants of these photons with a microwave antenna and kick off a revolution in cosmology.

As the Hubble expansion continued beyond this time, small clumps of matter began to grow through gravitational attraction amidst the ever-thinning gas. By the time the Universe was a billion years old, or about one-fifth the present age of the Earth, these clumps had spawned stars, galaxies, and clusters of galaxies. The young galaxies and clusters continued evolving and spreading farther apart as time elapsed, yielding the Universe we observe in the heavens today.

The visible Universe today has a temperature of $3°K$ and an average density of about a hundred billionths of a trillionth of a trillionth of a gram per cubic centimeter (10^{-31} g/cm^3). An earth-sized balloon full of this stuff would weigh less than a flea! Most of this matter is concentrated in clumps where the density is extremely high, relative to the universal average — in galaxies, stars, and planets. In between them are vast expanses of the most perfect vacuum imaginable, with hardly a single, solitary atom inhabiting every cubic meter.

———————————— • ● • ————————————

In addition to a beginning of time, Big-Bang cosmology permits an end of time. The Universe can expire, too. If there is enough matter around, the force of gravity will finally win out one day and rein in the seemingly relentless Hubble expansion. Everything will start collapsing back inward. Instead of the galaxies and clusters moving steadily apart, they will begin to approach one another. In such a case, the Universe will eventually become compressed to extremely high density and temperature — a condition that certain cosmologists have dubbed the "Big Crunch." Shortly thereafter, space and time as we know it will cease to exist. Fortunately, however, such a disaster could not occur for many billions of years, if ever.

If, on the other hand, the Universe has too little matter, then gravity will never be able to halt the Hubble expansion, which will go on literally *forever*. Space will keep on stretching, and time will never end. Matter will be spread ever more sparsely, and the average temperature of the Universe will fall steadily toward absolute zero. Eventually all the stars will exhaust their fuel supplies and die out. The Universe will finally become an extremely cold, lifeless realm populated only by gas, dust, and the dark cinders of dwarf stars. We call this dismal scenario the "Big Chill." Again, it could not happen for many billions of years.

We can liken these two scenarios to what happens when a rocket is launched. If the rocket has too little power to reach "escape velocity," something close to 7 miles per second, it falls back to Earth and crashes — as does an intercontinental ballistic missile. If it has enough power, on the other hand, it can escape the Earth's gravitational pull and will continue soaring out into interstellar space, as did the Voyager interplanetary probes after they flew by the outer planets. For both the rocket and our Universe, whether things fall back or not is merely a question of balance between the outward velocity and the inward pull of gravity.

Between these two extreme cases, however, there is a middle ground. If the rocket attains the escape velocity *exactly* (and has an appropriate trajectory), it will orbit the Earth as a satellite. Similarly, if an exact balance is struck between the Hubble expansion of the galaxies and their inward gravitational pull on one another, the Universe will continue to expand forever, but at a slower rate — never quite turning the corner. In such a case the Big-Chill scenario still occurs, but just barely.

The boundary line between these two scenarios, the Big Crunch and the Big Chill, is marked by what cosmologists call the "critical density." This is the average density of matter needed, about 5 millionths of a trillionth of a trillionth of a gram per cubic centimeter (or 5×10^{-30} g/cm^3), to arrest the expan-

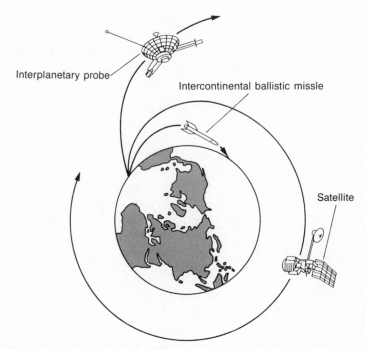

Three possible outcomes of a rocket launching. Depending on the velocity it attains, the rocket can fall back to Earth, launch a satellite into orbit, or send an interplanetary probe into outer space.

sion of the Universe but never quite turn it around. This is very little matter—about three hydrogen atoms in every 100 cubic meter. It's the density you would achieve if you somehow vaporized this page and then distributed its atoms throughout a volume a hundred times larger than the Earth's. If the average density of the Universe today is *greater* than this critical value, then gravity is strong enough to force the Big Crunch. If it is less than or equal to the critical density, we get the Big Chill.

Cosmologists use an important parameter, written as Ω (and called "omega," the last letter of the Greek alphabet), to describe the density of the Universe. Omega is the ratio of the actual density to the critical value. If it turns out to be greater than 1, the Universe will eventually end in the Big Crunch; if it is less than or equal to 1, we get the Big Chill. In the special case where Ω equals 1 exactly, we say the Universe is a "critical" universe. The value of omega determines the fate of the Universe.

According to Einstein's theory of general relativity, in which gravity is described as being due to the curvature of space itself, we can relate the various options for Ω to the overall curvature of the Universe. Of course it's difficult to perceive this very slight curvature, just as it's hard to tell that the Earth's surface is curved when you are standing in an Illinois cornfield. To detect the curvature of the Universe requires making extremely precise measurements over tremendous distances (billions of light-years). To date, such measurements are inconclusive.

A universe with Ω greater than 1 has the geometry of a closed, three-dimensional surface of a four-dimensional sphere. To help visualize this, consider a two-dimensional analogy. Think of this universe as an expanding spherical balloon in ordinary space, with clusters of galaxies sprinkled as dots upon its surface. A horizon in this picture is a circle drawn upon the surface of the balloon around a given dot. Such a "closed" universe contains only a finite amount of space, all of which will eventually be contained within any expanding horizon. In a closed universe, space is curved such that the sum of the angles of a triangle is *greater* than the 180 degrees expected in familiar Euclidean geometry. This is just what you discover if you draw a triangle upon the surface of a balloon and add up its angles.

A universe with Ω less than 1 has what scientists call a "hyperbolic" geometry that extends outward forever and contains an infinite amount of space. Here, the best two-dimensional analogy is a saddle-shaped surface with edges extending to infinity and all points having the same spatial properties as the saddle point itself. The sum of the angles of a triangle in a hyperbolic space is always *less* than 180 degrees. In such an "open," infinite universe, an expanding horizon can never encompass everything, no matter how long you wait.

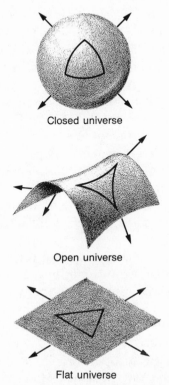

Closed universe

Open universe

Flat universe

Three possible geometries of the Universe: spherical, hyperbolic, and flat. In the first case the Universe is closed, corresponding to Ω greater than 1; it is open in the latter two cases, corresponding to Ω less than or equal to 1.

Between the open and closed universes there is a universe with Ω equal to 1 — a universe having exactly the critical density. The geometry in such a universe is the flat, Euclidean geometry most of us learned in high school — in which two parallel lines never intersect and the angles of a triangle add up to exactly 180 degrees. Here, the two-dimensional analogy is a flat plane extending to infinity in all directions. As in the hyperbolic case, such a flat universe implies an infinite amount of space, so this is a special example of an open universe. A very special example, as we shall see.

The modern ideas of open and closed universes originated in the work of the Russian meteorologist Alexander Friedmann. During World War I he had served as a balloonist, surveying the effects of aerial bombing on the Eastern Front. After the war he became interested in Einstein's theory of general relativity and began applying its equations to the dynamics of the Universe. In 1922, he published an article titled "On the Curvature of Space," which described a universe that exploded from a point in spacetime and ended in a great collapse — what we recognize today as a closed universe. But this paper appeared 7 years before Hubble's famous discovery, and nobody was very interested yet in such a dynamic, changing universe. Even Einstein, then enamored of a static, unchanging universe, thought the idea "suspicious."

Two years later Friedmann proposed yet another wild idea — an open universe that would continue expanding forever. He published this idea, too, but again nobody was listening. Friedmann unfortunately died of typhoid fever in 1925, before his mathematical insights about the Universe were vindicated by Hubble four years later.

Whether we live in an open or closed universe remains an unresolved question at present. Most cosmologists today strongly prefer the special case of an open universe with a flat geometry, but observations are as yet unable to verify this prediction convincingly. Such a flat universe, with Ω exactly equal to 1, seems necessary if we wish to understand how it has managed to continue expanding for billions of years without yet reaching a Big Crunch or Big Chill. And if Alan Guth's inflationary scenario is correct, Ω must equal 1 exactly. But modern science is based on observation and experiment, not on theoretical predispositions. What do we know for sure?

If we just count up the galaxies in a large volume of space, multiply this number by the average mass of visible stars in a galaxy, and then divide by the total volume, we should be able to estimate the average density of visible matter. But the result of such an estimate is only about one-half percent of the critical density, or an Ω of 0.005. The detailed measurements used in this estimate could be off by enough to make Ω twice as large, but no more. At most, then, luminous matter can contribute only 1 percent of what is needed to reach the critical density. If the inflation idea is true, that is, there has to be at least a *hundred times* more matter that we cannot see.

By examining the motions of the stars and gas within nearby galaxies, as is done at length in Chapter 3, we can conclude that there must be a lot more unseen matter in and around them, maybe ten times what is visible in our telescopes. Perhaps clouds of gas or dust, planets the size of Jupiter, dark stars, or even black holes are making the dominant contribution to a galaxy's total mass, not the luminous stars. Even counting the unseen mass that seems to be associated with galaxies, however, we only achieve an Ω of around 0.1 — still well short of the critical density.

Turning next to the synthesis of light elements during the Big Bang, the subject of Chapter 4, we conclude that ordinary, garden-variety matter familiar to us from everyday life can contribute at most 15 percent of the critical density, or Ω equals 0.15. Any higher, and there would be different amounts of light elements than what we observe in the Universe today. This limiting value is roughly consistent with the density estimated above from the motions of galaxies, but is completely inconsistent with an Ω of 1.

To reach the critical density requires not only that the great majority of matter in the Universe be dark; it must also be some new and strange kind of matter very different from the familiar electrons, protons, and neutrons making up our bodies and everything around us. And the great bulk of this mysterious dark matter cannot even collect near the visible aggregations, or else we would already have detected its existence indirectly — through its gravitational influence on their motions.

What this dark matter is, and exactly where it is hiding, is one of the deepest mysteries confronting science today.

2

The Particle Connection

*E*arly in this century, while Einstein, Friedmann, and Hubble were expanding humanity's horizons to encompass the entire Universe, other scientists were delving *inward*, seeking to unearth the submicroscopic structure of matter. They were predominantly Europeans, sporting august-sounding names like J. J. Thomson, Ernest Rutherford, Niels Bohr, Werner Heisenberg, Erwin Schrödinger, and James Chadwick. Between 1895 and 1930, these men and others pried apart the atom and discovered a new realm of "subatomic" particles swarming around inside. In the process they also developed the theory of quantum mechanics, one of the bastions of modern physics. Little did they suspect, however, that their investigations of the smallest objects in existence would one day lead to a deeper understanding of some of the biggest structures in the Universe.

Originally proposed by the ancient Greek philosophers Leucippus and Democritus, the "indivisible" atoms of which ordinary matter is composed had been placed on a solid empirical foundation only during the latter part of the nineteenth century. But scientists quickly showed that these atoms are not really indivisible or "elementary," after all. Instead they are built from far tinier fragments. The redoubtable British physicists J. J. Thomson and Ernest Rutherford led the way with two key experiments in 1895 and 1910. Inside every atom, they discovered, there are wispy, negatively charged particles called "electrons" gyrating around a very small, positively charged core thousands of times heavier, known as its "nucleus."

This new view of matter had barely taken hold, however, when Rutherford and his Cambridge University colleague James Chadwick showed that an atomic nucleus was *itself* built of tiny fragments. Hardly a trillionth of an inch across, a nucleus can be broken down into positively charged particles called "protons" and electrically neutral particles known as "neutrons." A hydrogen atom, for example, has only a single speedy electron whirling around a nucleus consisting of a lone proton. A normal oxygen atom has eight electrons orbiting about a nucleus with

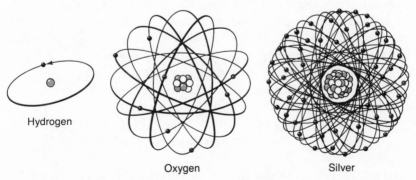

Hydrogen

Oxygen   Silver

Atoms of hydrogen, oxygen, and silver. Their atomic nuclei, composed of protons and neutrons, are surrounded by a cloud of electrons; equal numbers of electrons and protons in a given atom make it electrically neutral.

eight protons and eight neutrons inside. Uranium, the heaviest atom that occurs naturally, has a veritable swarm of 92 electrons surrounding a nucleus built of 92 protons and at least 140 neutrons.

To get some idea of the relative sizes involved, try to imagine that an atom is somehow enlarged by a factor of a trillion. Then it would be nearly the dimensions of a football stadium. The electrons would be smaller than tiny fleas jumping about in the stands. And the nucleus at the center of the atom would be hardly the size of a housefly buzzing around near the 50-yard line. The vast majority of an atom is empty space; more than 99.9 percent of its entire mass is concentrated in its minuscule core!

By far the dominant contribution to the mass of all ordinary matter, which is composed of atoms, is concentrated in the protons and neutrons of their atomic nuclei. A proton weighs 1836 times as much as an electron, and a neutron a bit more, 1839 times. Both are members of a class of particles physicists call "baryons," a name derived from the Greek word *barys*, meaning "heavy." The electron, on the other hand, belongs to a different class of particles known as "leptons," from the Greek *leptos* for "small." All the matter that we can see is made of baryons and leptons.

What keeps the electrons whirling around a nucleus, instead of flying off in all directions, is an attractive force known as the electromagnetic force — the same force responsible for lightning and electric current. Just as massive bodies exert gravitational force, which keeps moons orbiting planets and the planets orbiting the Sun, charged particles exert electric and magnetic forces upon one another. Positively charged protons in the nucleus attract the negatively charged electrons to it, holding them close by and rendering the atom electrically neutral as a whole.

Actually, the electrons do not "orbit" the nucleus along well-defined paths such as those followed by planets orbiting the Sun. After Rutherford's 1910 discovery of the nucleus, the young Danish theorist Niels Bohr tried to concoct such a "planetary" model of the atom, with its electrons relegated to a very specific group of orbits. This Bohr–Rutherford model worked fairly well for the simplest atom, hydrogen, but was unable to explain the behavior of more complex atoms like helium, which have additional electrons speeding about inside.

The ultimate resolution of these problems came with the development of quantum mechanics in the mid-1920s. First the brilliant German theorist Werner Heisenberg recognized that it is impossible to determine both the position and the momentum of a subatomic particle *exactly* at any specific instant. There is always an irreducible amount of uncertainty that enters the two measurements, which he expressed in his famous "uncertainty principle." Following a different but equivalent path, the Austrian physicist Erwin Schrödinger introduced the idea of a particle's "wave function," which tells the chances of finding that particle at a chosen place and time.

Thus, the appropriate way to think of an atom is to regard it as a fog, or cloud, of electrons surrounding its central nucleus. The cloud represents the wave function of these electrons, which determines our probability of finding an electron should we ever try to make a measurement at a specific point. Inside the cloud there are various "energy levels" corresponding to the different energies of the electrons, which can assume only a specific set of definite values—as in the old Bohr–Rutherford model of the atom.

By applying energy to an atom, say by heating it, we can kick one of its electrons momentarily up into a more energetic level. Left to itself thereafter, the atom quickly returns to its preferred "ground state." The electron reverts to its original, lower energy level, and the atom emits a particle of light, or photon. It is these photons, striking the retinas in our eyes, that allow us to see objects around us and avoid bumping into them.

Something like this process happens when we heat an iron poker in a fireplace: the iron atoms become excited and begin emitting photons. If we heat the poker long enough, it begins to glow—first a dull red, then yellow as it gets hotter, and finally a bright white as it becomes extremely hot. The longer wavelength red light is due to streams of lower energy photons hitting our eyes; shorter wavelength yellow and white light results from photons of higher energies, which are emitted more copiously as the poker's temperature rises.

Fired by intense thermonuclear processes blistering its central core, the Sun produces light in a similar fashion. So do the myriads of stars seen at night. Heat percolates slowly outward from the nuclear furnaces at the centers of these stars, maintaining their luminous surfaces at thousands of degrees. Cooler stars have a reddish tinge, and the very hottest are bluish or pure white. An average star like our Sun sports a yellowish hue, evidence of a moderate surface temperature.

All the light that pervades our daily existence originates in atomic or molecular processes such as those described above, whereby atoms or their electrons are jostled and emit visible photons. Invisible photons are often emitted, too — in the infrared and ultraviolet parts of the electromagnetic spectrum. Very low energy photons, also known as "radio waves" and "microwaves," emerge whenever electrons are jiggled inside an antenna or cavity. Photons of many kinds and hues originate because the atoms of normal matter are continually undergoing transformations from one state of matter to another.

These photons are the principal means by which we recognize that matter exists. Here on the Earth's surface, our sense of sight is aided by those of hearing, touch, taste, and smell. But to detect the presence of matter beyond our planet, we can rely only on what we can see (visible light) and the other forms of electromagnetic radiation that sophisticated instruments enable us to discern.

In quantum mechanics, the electromagnetic force keeping electrons near atomic nuclei can be thought of as occurring through the *exchange* of photons. In this picture, an electron emits a photon, which is subsequently absorbed by one of the protons inside the nucleus, or vice versa. Swapping photons like this, again and again, the electron and nucleus remain in close contact with one another—two intimate trading partners in the subatomic economy. Photons are the "carriers" of the electromagnetic force that glues atoms together.

This is not the only force, however, operating on subatomic particles. After Chadwick and Rutherford showed the atomic nucleus to be composed of protons and neutrons, some kind of very strong force was needed to bind them together. Otherwise, left to themselves, all the positively charged protons in a nucleus would repel one another, and the nucleus would simply explode. As you know, like charges repel. Therefore there had to be *something* that kept them packed together inside such a tiny globule. In 1935, the Japanese theorist Hideki Yukawa predicted that such a strong force was carried by a new kind of particle, later dubbed the "meson" (pronounced MEZ-on, from the Greek word *mesos* for "middle"), which had to be about 200 times as heavy as an electron.

When physicists began searching for mesons in the showers of cosmic rays raining down from the heavens, however, they discovered a plethora of hitherto unexpected new particles. At 264 times the electron mass, the "pion" (pronounced PIE-on) fitted Yukawa's prediction. Flitting back and forth between the protons and neutrons, pions can indeed bind an atomic nucleus together very tightly. But what were all the other particles—the so-called muons, kaons, lambda, sigmas, and xis—that also turned up in particle detectors?

Lacking a satisfactory theory to explain all these new particles, the physicists of the 1950s began classifying them according to their intrinsic properties—mass, charge, spin, and other, less familiar characteristics. The "muon" (pronounced MEW-on) was found to be a lepton like the electron, although it is about 200 times as massive; neither feels any effects of the strong nuclear force. The lambda, sigmas, and xis were classified as baryons like the proton and neutron because they were heavier particles that *do* feel the strong force. Sporting intermediate masses, the pion and kaon (pronounced KAY-on) fall into the group of particles known as mesons—the name originally coined for Yukawa's carrier of the strong force.

Theoretical and experimental advances of the 1960s proved, however, that mesons and baryons are *not* elementary particles, after all. Instead they are built from still tinier entities called "quarks" (pronounced KWARKS), predicted in 1964 by two theorists, Murray Gell-Mann and George Zweig of the California Institute of Technology (CalTech). An urbane and literate polymath with a knack for inventing new words, Gell-Mann coined the whimsical name for these particles from a line in James Joyce's novel *Finnegans Wake*, "Three quarks for Muster Mark!"

Not until the late 1960s, however, did convincing evidence for quarks begin to appear in experiments performed at the Stanford Linear Accelerator Center (SLAC). A group of physicists from the Massachusetts Institute of Technology (MIT) and SLAC, led by Jerome Friedman, Henry Kendall, and Richard Taylor, bombarded protons and neutrons with high-energy electrons supplied by the accelerator. A fraction of these electrons were seen to ricochet off at large angles—as if they had struck something hard and tiny inside. Subsequent experiments performed during the 1970s at SLAC and other accelerators all around the world confirmed that quarks are indeed the fundamental building blocks of all nuclear matter.

In the original theory proposed by Gell-Mann and Zweig, there were three different kinds of quarks (which is one reason he chose their name)—dubbed "up" or *u*, "down" or *d*, and "strange" or *s* quarks. Unlike the proton and electron, which have whole-number values of electric charge (+1 and −1), these quarks have *fractional* charges of $\frac{2}{3}$ and $-\frac{1}{3}$. Because fractional charges had never been witnessed in nature, most physicists (including Gell-Mann himself) at first had great difficulty taking the idea seriously. It was the MIT–SLAC experiments, which detected quarks flying around inside protons and neutrons, that encouraged physicists to accept them as *real* parti-

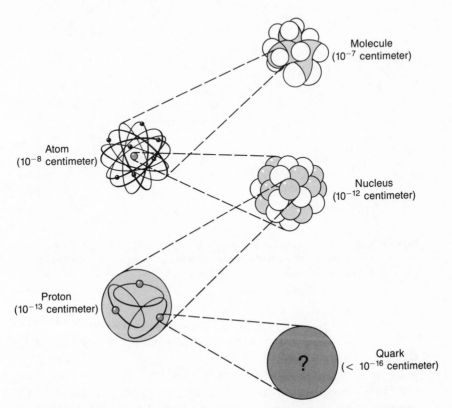

The composite structure of ordinary matter. Molecules are built of atoms, which contain nuclei made of protons and neutrons that are themselves composed of quarks held together by gluons. As far as we know today, quarks have no internal structure, but physicists have made measurements only down to the level of 10^{-16} centimeters.

cles. A mountain of indirect evidence now attests to their existence.

Baryons are built from three individual quarks, and mesons are composed of a quark plus an "antiquark" — the antimatter counterpart of a quark. The proton, for example, is made of two up quarks plus a down quark; the neutron contains two down quarks and an up quark. Pions are composed of an up quark plus

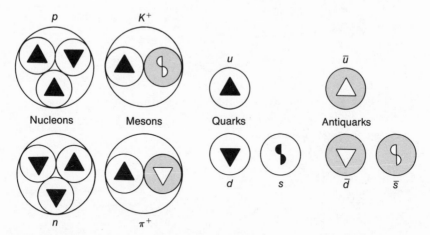

Baryons and mesons are made of quarks. Baryons such as the proton (*p*) and neutron (*n*) contain three quarks, while mesons (π^+ and K^+) are composed of a quark plus an antiquark.

an antidown quark, or a down quark plus an antiup quark. The so-called strange mesons and baryons contain at least a single strange or antistrange quark.

As far as we know today, leptons and quarks are the ultimate building blocks of ordinary matter. They are incredibly tiny objects, much less than a millionth of a billionth of a centimeter across. In an atom the size of a football stadium, quarks would still be too small to see. For all we know, they could even be mathematical points with absolutely no size at all!

What is remarkable about these quarks is the apparent fact that they never appear alone in nature. The most sociable of particles, they come only in pairs or trios. No physicist has ever witnessed a solitary quark in any detector — although a few have *thought* they had. The force binding quarks together inside mesons and baryons is apparently so incredibly strong that it never lets go.

Just as with the forces binding electrons to nuclei, and protons and neutrons within nuclei, this interquark force is also borne by particles, called "gluons" (pronounced GLUE-ons). The gluons flit back and forth between quarks, imprisoning them forever within the confines of mesons and baryons. The gluons themselves are locked up inside, too, and never appear in nature all by themselves.

This coherent picture of the subatomic world is part of what is called the "Standard Model" of elementary-particle physics. It emerged during the 1960s and 1970s, at about the same time as the revolution that led to the standard Big-Bang cosmology discussed in Chapter 1. The Standard Model holds that all matter (or at least the ordinary stuff we are familiar with) is composed of leptons and quarks. The forces between these building blocks are carried by other elementary particles, such as the photons and gluons, which are known collectively by the rather technical term "gauge bosons."

The Standard Model prescribes how subatomic particles behave at high energy and when they get extremely close together. These are just the conditions that prevailed during the first second of the Universe's existence — when the temperature was more than 10 billion degrees. Particles then were racing about at tremendous velocities close to the speed of light, bashing into each other repeatedly and converting from one kind to another.

The Standard Model has proved an invaluable tool for physicists trying to reconstruct the exact details of what happened, on the submicroscopic level, during the early moments of the Big Bang. Studies of matter under the extremely violent conditions that are reproduced almost daily in earthbound particle accelerators have generated a solid foundation of established facts upon which to base cosmological arguments about the origins of existence.

———————————— • ● • ————————————

Besides the electron and muon, there are other kinds of leptons in the Standard Model. "Neutrinos" (pronounced new-TREE-nose) are neutral, extremely light particles that interact very weakly with matter. They were conceived in 1930 by the

Austrian theorist Wolfgang Pauli, who suggested that such a particle might be responsible for spiriting off the large amounts of energy that seemed to be missing after certain radioactive decays of atomic nuclei. A few years later the Italian physicist Enrico Fermi (who actually named the neutrino) formulated a theory explaining these decays in terms of the emission of these ghostly particles.

It was more than two decades, however, before neutrinos were finally discovered. So feeble is their interaction with matter that they can easily penetrate millions of miles of lead without ever stopping. They can sail right through the vast empty spaces inside an atom and never disturb it at all. Although there was plenty of indirect evidence for their existence, they were not observed directly until 1956, when Clyde Cowan and Frederick Reines of the Los Alamos National Laboratory detected neutrinos spewing forth from a nuclear reactor in South Carolina.

There are a few different kinds of neutrinos, too, each associated with a particular charged lepton. An "electron neutrino" is closely affiliated with the electron, and a "muon neutrino" with the muon. In 1962, Leon Lederman, Melvin Schwartz, and Jack Steinberger of Columbia University did an experiment at the Brookhaven National Laboratory showing that whenever a charged pion decays, it emits a muon and a muon neutrino—but never an electron neutrino. When an atomic nucleus decays, by contrast, it emits an electron and an electron neutrino (or their antiparticles), but never a muon neutrino. In the Standard Model, leptons always come in pairs.

As far as we know, these neutrinos are stable elementary particles that do not disintegrate. They can emerge from radioactive decays of atomic nuclei, but are not constituents of normal matter—at least not in the usual sense. There are vast swarms of them speeding throughout the Universe—remnants of the Big Bang that can never settle down to a humdrum existence inside an ordinary atom. Neutrinos are the vagabonds of subatomic society.

The only way neutrinos can interact with normal matter is through an extremely feeble force known as the "weak force," which is thousands of times feebler than the electromagnetic force. The weak force only extends over distances less than a millionth of a billionth of a centimeter, or about a hundredth the size of a proton. That is why neutrinos are such penetrating particles. They interact only when they come within this tiny distance of other leptons or quarks—and that happens very infrequently.

Having read this far, you might now suspect that the weak force is carried by a particle, too, and you would be essentially correct. Actually it is carried by not one but two gauge bosons known as the "W" and "Z" particles. They were predicted by theoretical physicists in the 1950s and 1960s, and finally discovered in 1983 by a large international collaboration led by the Italian physicist Carlo Rubbia. The W carries an electric charge of either +1 or −1; the Z, however, is electrically neutral like the photon and gluon. Both are massive in the extreme—almost a *hundred* times heavier than the proton.

If you shoot a neutrino across the bow of an electron or quark, it only interacts with them by swapping W or Z particles back and forth. It cannot exchange photons or gluons. Because the W and Z are so terribly ponderous, however, the neutrino must get very close for the exchange to occur. It's a bit like two players tossing balls back and forth. If they throw a light tennis ball, they can roam far afield and reach one another with no difficulty. But to pass a heavy medicine ball they must approach within a few feet and practically *hand* it from one to the other.

These W and Z particles, however, can never be found lying around in ordinary matter. They are far too heavy, and have only the most fleeting existence as carriers of the weak force. To create them, Rubbia and colleagues had to concentrate tremendous amounts of energy in a very tiny volume. They accomplished this by smashing high-energy protons into their antiparticles (antiprotons) using a huge particle collider 4 miles in circumference at the European nuclear research facility CERN

Aerial photograph of the CERN laboratory and surrounding regions near Geneva, Switzerland. The W and Z particles were discovered in 1983 at the circular collider labeled SPS. Since 1989, Z's have been produced in quantity at the larger ring, known as LEP.

near Geneva, Switzerland. Once produced, these massive W and Z particles disintegrate almost instantaneously into pairs of quarks or leptons — plus lots of energy. They survive less than a trillionth of a trillionth of a second.

Today colliders that smash electrons into positrons (their antimatter counterparts) produce copious quantities of Z particles for detailed studies of their properties. A big circular collider 17 miles in circumference, the Large Electron–Positron collider known as LEP, began operations at CERN in 1989, as did a smaller, 3-mile-long machine at SLAC called the Stanford Linear Collider (SLC). The LEP collider should eventually be able to produce large numbers of W particles, too.

The last time there were so many W and Z particles in existence was during the earliest moments of the Big Bang. During the first trillionth of a second, temperatures were extremely

Aerial photograph of SLAC. Quarks were discovered in the large concrete building at the center, near the end of the 2-mile linear accelerator; this device now supplies high-energy electrons and positrons to the hall at lower right, where they annihilate each other, producing Z particles.

high and particles were flying around at energies even greater than those produced by LEP and SLC. When two such particles smashed into one another, which happened very frequently because everything was packed so closely together, they often produced a W or Z particle. By generating these massive particles in quantity for the first time in over 10 billion years, modern particle colliders are now providing scientists a glimpse of conditions that existed at the very birth of the Universe.

• ● •

There are several other elementary particles in the Standard Model; most of them are unstable like the W and Z. They were discovered during the 1970s at particle accelerators and colliders in the United States. SPEAR, a circular electron – positron collider at SLAC, was especially productive in this regard. A third lepton, the "tau" particle (τ), and a fourth quark, the "charm" quark, turned up originally at this machine.

In the Standard Model, quarks are supposed to occur in pairs like the leptons. But in the early 1970s there were only three known quarks — up, down, and strange. Physicists could pair the up and down quarks together, but the strange quark had no obvious mate. Some theorists therefore began to suggest there should be a fourth quark lying around undiscovered, and a few experimenters set out to search for it.

The first evidence for its existence came somewhat by surprise in late 1974, when physicists led by Burton Richter of SLAC discovered a meson they called the "psi" (pronounced SIGH), or ψ, particle at SPEAR. At about the same time another team led by MIT physicist Samuel Ting found the very same meson at Brookhaven, and dubbed it the J particle. Like other mesons, the J/psi (as it has come to be known) is not elementary; it is made up of a charm quark, written as c, paired with its antiquark.

The tau particle is a charged lepton with about twice the mass of the proton. Discovered at SPEAR in 1976 by SLAC physicist Martin Perl and his colleagues, it survives less than a trillionth of a second. Except for its mass, which is over 3 thousand times greater, the tau particle is identical to an electron. Like the electron and muon, the tau has a neutrino paired with it, called the "tau neutrino." The evidence for the existence of this neutrino is strong, but so far remains indirect.

The quarks and leptons of the Standard Model are organized into what physicists call generations or families, each consisting of a pair of quarks and a pair of leptons. All ordinary matter is built up from members of the first family — up and down quarks plus the electron and its neutrino. The second family contains the strange and charm quarks plus the muon and muon neutrino.

Continuing this pattern, one would expect a *third* pair of quarks to fill out a third quark–lepton family, whose leptons are the tau and its neutrino. Indeed, the first member of this pair—

Aerial photograph of Fermilab, where the first evidence for the bottom quark was discovered in 1977. Protons and antiprotons travel in opposite directions inside the 4-mile ring, producing the highest energy collisions possible today.

a massive quark dubbed "bottom" or b, more than five times as heavy as a proton — was discovered in 1977. A team of scientists led by Leon Lederman found the first evidence for the existence of the bottom quark at the Fermi National Accelerator Laboratory (Fermilab) west of Chicago. Since that time physicists around the world have been eagerly searching for its anticipated mate, dubbed the "top" or t quark, but so far without success. All that can be said to date is that the top quark is at least 95 times as heavy as the proton.

Today these exotic quarks and leptons are produced only at particle accelerators or in cosmic rays, but during the earliest moments of the Big Bang they were as common as the particles of the first family. They would have been produced quite abundantly during the first billionth of a second, when particle energies were sufficiently high. Most of them are unstable, however, with lifetimes ranging from a few millionths to a few trillionths of a second. Only the neutrinos seem to survive a long time.

To make a substantial contribution to the mass of the Universe today, a particle must be stable (or be bound up in a composite

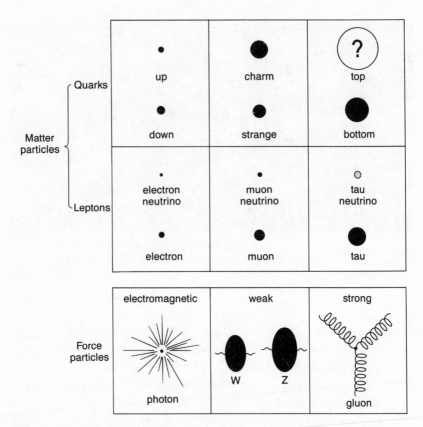

The known particles of the Standard Model. Matter particles are grouped in families, each consisting of two quarks and two leptons. (The more massive a particle, the larger it is drawn above, but none of these particles have a detectable size.) The top quark and tau neutrino have not yet been officially "discovered," but there is strong evidence that they do indeed exist.

particle that is stable). If a certain kind of quark or lepton disintegrates rapidly, it could never have survived the billions of years that have elapsed since the Big Bang, and therefore it cannot be part of the dark matter that now seems to exist. Thus, a necessary characteristic of any dark-matter candidate is an extremely long lifetime — much longer than 10 billion years.

Electrons and protons are stable particles, as are neutrons when they are bound together with protons in such stable nuclei as that of the helium atom—which is composed of two protons and two neutrons. Of all the remaining particles in the Standard Model, however, only the humble neutrinos are stable enough to make a contribution to the mass of the Universe, and then only if they carry some intrinsic mass of their own. At present, this is an open question. For all we know, neutrinos may have no mass at all.

———————————— • ● • ————————————

The Standard Model of particle physics has proved to be extremely successful in describing the behavior of elementary particles. How quarks, leptons, and gauge bosons interact with one another at high energies, or when they get extremely close, can be readily predicted using its equations—and the principles of relativity and quantum mechanics. The Standard Model can account for all subatomic phenomena known today.

But the seeming complexity of this supposedly fundamental theory is exceedingly troublesome to physicists. We have a common belief, verging on religious faith, in the basic *simplicity* of nature. We would prefer to have explained everything using only a few elementary building blocks, but in the Standard Model there are at least 12—plus 4 particles responsible for carrying forces. This is altogether too many.

Such a state of affairs has driven many physicists to search for a deeper, simpler theory that would encompass the Standard Model but reach beyond it. This yearning for simplicity is a common theme in twentieth-century physics. Whenever the full collection of supposedly "elementary" entities has gotten too unwieldy, as became the case with atoms and occurred again with nuclei, physicists have sought a deeper explanation. This is the situation we find ourselves in today.

There are many attempts to go beyond the Standard Model and develop theories that explain the observed patterns of quarks, leptons, and gauge bosons. Why do they have the various masses they exhibit? Are the different forces between these particles just different manifestations of each other — just as ice, water, and steam are different forms of H_2O? These questions cannot be answered within the framework of the Standard Model itself.

In many attempts to peer beyond the Standard Model, physicists take their cue from the unification of the weak and electromagnetic forces that occurred during the 1960s. Previously thought to be completely disparate phenomena, these two forces were shown to be one and the same through the revolutionary ideas of theorists Sheldon Glashow, Abdus Salam, and Steven Weinberg. Glashow was the first of them to work on this problem; in 1960 he predicted the Z particle as an additional carrier of the weak force. Later in that decade Salam and Weinberg discovered how to combine the two forces into one while generating large masses for the W and Z particles. The 1983 discovery of the W and Z particles confirmed these ideas beyond any doubt.

That unification of the weak and electromagnetic forces has spawned numerous efforts to bring the strong nuclear force, and even gravity itself, into a single theory that explains all these forces. Such ideas as grand unification, supersymmetry, or the superstring theories are touted as the wave of the future. Some physicists today claim that a so-called Theory of Everything has even been formulated already, and only awaits experimental confirmation. These are heady times in high-energy physics.

These unified theories have some features that make them extremely appealing to cosmologists. For one, their effects would have been manifest only during the earliest fraction of a picosecond (a trillionth of a second) of the Big Bang, when average particle energies had to be so unspeakably high that any differences between the individual forces simply evaporated. It was only after the Universe cooled down, according to this thinking, that these differences became apparent.

Another common feature of theories that reach beyond the Standard Model is the prediction of new, as yet unobserved particles. One hears talk of axions, squarks, photinos, sleptons, gluinos, and a whole host of other hypothetical beasts. These

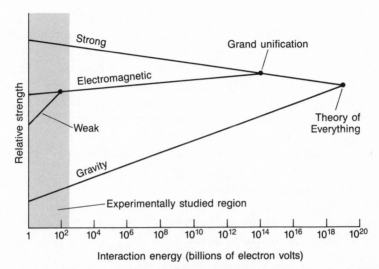

Unifications of the four fundamental forces of nature. At extremely high energies such as occurred during the early moments of the Big Bang, the different forces are believed to have been completely indistinguishable.

particles must be extremely massive or interact very feebly with normal matter (quarks and leptons); otherwise they would already have been detected at particle accelerators. If they were stable, such exotic new particles would be excellent candidates for the invisible content of the Universe, because they could have been produced copiously during the Big Bang and might have lasted until the present day. We would hardly know they existed, but they could make a major contribution to the mass of the Universe.

Unfortunately, however, the unified theories that require such ghosts come into play only at truly tremendous energies, which makes them all but impossible to test in an earthbound experiment. Using particle acceleration technologies available today, colliders capable of generating such energies would have to be bigger than the Solar System and perhaps even as large as the entire Milky Way. Though governments have been kind to high-energy physics recently, there are limits!

More and more, physicists are searching the heavens for an answer. The same unified theories that may have been dominant during the earliest split second of the Big Bang should also have left dim relics behind, dark fossils in a world beyond our immediate perception that help determine how the Universe is structured today. Through the force of gravity, which acts upon all forms of matter, these shadowy remnants can make their presence felt by their effect upon what we indeed *can* see. The shapes and motions of the galaxies—and the way they are sprinkled about in outer space—may hide crucial secrets about the earliest moments of existence.

By trying to locate and decipher these ghostly remnants, these shadows of creation, we may ultimately hope to understand the very birth of the Universe itself.

3

Shadows of Doubt

*D*uring the 1970s, astronomers began to recognize that there is far more to the Universe than meets our eyes. Like snowcaps on faraway mountaintops, the luminous stuff that can be seen in the night sky is but a small fraction of a much larger whole. Astronomers reached this startling conclusion by patiently recording the movements of stars and gas within galaxies and by studying how galaxies themselves speed about one another in clusters. Using little more than Newton's law of gravity and some careful observations, they discovered that there must be vast amounts of invisible matter out in deep space, making the glowing spots seen in telescopes move so surprisingly fast.

With the best observatory in the world at his disposal, the flamboyant Danish astronomer Tycho Brahe made the first detailed measurements of planetary motions late in the sixteenth century. Dying of a burst bladder after a long drinking bout, he bequeathed his treasured charts to Johannes Kepler, the mystical German mathematician who had been serving with him at the court of the Holy Roman Emperor in Prague. He wanted his colleague to use these data to disprove the heliocentric system of Nicolaus Copernicus, but Kepler instead employed them to prove and extend the Copernican ideas. In 1619, he made two crucial observations: the planets follow *elliptical*, not circular, orbits and move more slowly in these orbits the farther they are from the Sun.

In his 1687 masterwork, the *Principia Mathematica*, Isaac Newton drew upon the detailed groundwork laid by Brahe and Kepler to derive a mathematical law for the force of gravity

Isaac Newton. His mathematical law describing the force of gravity, published in 1687, explained the detailed motions of moons and planets.

between any two celestial bodies. It had to increase, he concluded, as the product of their two masses and to decrease as the square of the distance between them. This force, and not some imaginary crystal sphere or other mechanical construct, was what determined the motion of the Moon about the Earth. The exact same gravitational force, too, could easily explain Kepler's empirical laws of planetary motion.

These motions give clear evidence that the strength of gravity falls off rapidly with distance. We all realize that the Earth moves completely around the Sun in 1 year. The outer planets, say Jupiter or Pluto, take much longer — 12 years for Jupiter and 249 years for Pluto. And while Jupiter has farther to travel than Earth, its orbital path is only 5 times as long, not 12; Pluto's orbital path is 40 times Earth's, not 249. Therefore, Jupiter's orbital *speed* (which equals the distance traveled divided by the time elapsed) is substantially less than Earth's, and Pluto's is even smaller yet. Their slower velocity occurs because the force of gravity drops off substantially in the vicinity of the outer planets. Since the Sun carries the vast bulk of the Solar System's mass — the source of gravity — this force gets steadily weaker and weaker the farther out you go. In fact, it drops by a factor of 4 every time the distance from the Sun doubles.

Knowing the distances between any pair of gravitating bodies, and their velocities, one can work the other way around and employ the law of gravity to learn something about their masses. By watching how the Moon orbited the Earth, for example, Newton was able to estimate the Earth's mass and to show that the same gravitational force influenced moons, planets, and falling apples. Celestial and terrestrial motions, he was the first to recognize, are described by the very same universal law of gravity. Prior to his work, they were thought to be unrelated phenomena. In effect, Newton was the first person to unify two forces into a single one.

• ● •

Following in Newton's footsteps, modern astronomers can use his law of gravity to estimate the mass of our own galaxy, the Milky Way, from the motions of its component parts. The Sun, for example, whirls around the Milky Way at a distance of about 30 thousand light-years from the galactic center; its speed about this center is nearly 200 kilometers per second, or 450 thousand miles per hour. Knowing just these two simple facts, one can estimate the total mass of all the celestial objects lying within the orbit of the Sun; it turns out to be about 100 billion times the Sun's mass, or 100 billion solar masses in scientific terminology. Since the Sun is just an average star that cruises the galaxy out near the edge of its visible region, we might well conclude that there are about 100 billion suns inside the Milky Way. That's about as many as the number of grains in a shopping bag full of sand.

Astronomers make the same kinds of calculations for such nearby galaxies as the Andromeda galaxy, known also as M31. This fuzzy patch of light is the farthest object you can see with the unaided eye. It lies about 2 million light-years away. Together with the Milky Way, M31 forms what astronomers call a "binary pair" — two galaxies that gravitate about one another. An orderly spiral galaxy like our own, M31 also seems to have about a hundred billion suns within its luminous regions. This seems to be a typical size for spiral galaxies. (Galaxies come in three basic shapes — spirals like M31 and the Milky Way; ellipticals, where the stars are distributed in an egg-shaped fashion; and irregulars, which have no such orderly pattern.)

Applying this same kind of reasoning to the farthest reaches of space, astronomers can make a rough guess at the density of matter in the visible universe. They simply count up the average number of galaxies in a typical volume and multiply this number by the ordinary galactic mass of 100 billion suns. This is, however, an oversimplification of the way astronomers work. In actual practice, they "add up the light" coming from such a typical volume and then multiply this total by a "mass-to-light ratio" that reflects how much mass is associated with the ob-

The Andromeda galaxy, or M31. This is the nearest spiral galaxy to our own Milky Way. The two galaxies orbit one another as a binary pair, and are not moving apart with the general Hubble expansion.

served emissions. This approach yields roughly the same result as counting up the number of galaxies.

Because there are important ambiguities in these calculations, the density so estimated is accurate only to a factor of about 2. But such a "visible density" still turns out to be far less than the critical density we discussed in Chapter 1. Give or take a factor of 2, the density of visible matter in the Universe is a mere 0.5 percent of the all-important critical value, or $\Omega = 0.005$. If there were no other matter than what can be detected with telescopes or radio antennas, our Universe would be an extremely open universe indeed, expanding forever to fizzle out eventually in the Big Chill.

But there is something puzzling about these observations. If we look at our own galaxy, for example, and try to add up all the various masses within it, we fall well short of our first estimate of its total mass, which was based on the Sun's motion. Even after adding in all the gas and dust, we still come up short. We get less than half the apparent mass inferred from stellar dynamics. Following earlier work done by Jan Oort of Holland, John Bahcall, an astrophysicist at Princeton's famed Institute for Advanced Study, rediscovered this odd discrepancy during the 1970s, and his conclusions have been confirmed repeatedly. Either Isaac Newton was wrong, or more than half the mass of the Milky Way is invisible.

If we now glance over at our neighboring galaxy M31, we notice a similar anomaly. As we look out beyond the outer fringe of its spiral disk, beyond the great bulk of its visible matter, we notice that the speeds of the globular clusters and dwarf galaxies darting around it do *not* decrease with distance as expected from Newton's law. Instead the speeds of these stellar swarms level off at a roughly constant value. So does the rotational speed of the clouds of hydrogen gas that are found well beyond the visible fringe of M31. This is strange.

Remember that the planetary speeds in the solar system fall off steadily the farther a planet is from the Sun. A similar thing should happen with the rotational speeds in M31 the farther one travels from the galactic center — if most of the mass is truly concentrated there. If there is instead plenty of extra, unseen matter that extends *beyond* the visible disk, however, then these speeds can remain relatively high at such distances.

Astronomers measure velocities of distant objects by studying the light coming from them and seeing how the characteristic wavelengths (or atomic lines) emitted by hydrogen shift relative to their normal positions in the electromagnetic spectrum. If these emissions shift to longer wavelengths, toward the red end of the spectrum, the object in question is moving away; if they shift to shorter wavelengths, toward the blue end, the object must be moving in our direction. The amount of this "Doppler shift" determines the speed of the object. In a similar fashion the apparent frequency, or pitch, of a train whistle is higher (corresponding to shorter wavelengths of the sound waves) when the train is approaching and lower (longer wavelengths) when it is receding from a listener.

As Hubble first noticed during the 1920s (see Chapter 1), virtually all galaxies reveal redshifts in their light, which means they are receding from us. M31 is one of the few galaxies that instead exhibits a "blueshift." Under the influence of gravity, it is moving *toward* the Milky Way. Thus, the wavelengths of the light emitted by its stars are shifted toward the blue end of the spectrum.

To estimate the relative speed of two distant objects, say the center of M31 and one of its satellites, astronomers measure the *difference* between their redshifts or blueshifts. This is in fact how they determined that the speeds of these satellites do not drop off beyond the edge of M31's disk, but seem instead to flatten out at a constant value.

Such "flat rotation curves," as they are known in the jargon, began to be noticed by astronomers in the 1970s. A rotation curve is a graph showing the rotation speed versus distance from the galactic center; if the measured speeds level off at large distances, the curve is said to go flat. Astronomer Vera Rubin and her associates at the Carnegie Institution in Washington, D.C., first measured such a flat rotation curve for M31 in the early 1970s, but they did not make very much of it at the time.

When more and more flat curves showed up, however, they began to get suspicious and published their discovery. By the end of the decade, the evidence had become fairly conclusive: all spiral galaxies have flat rotation curves. From these curves astronomers concluded that luminous matter makes up no more than half, and probably less than a fourth, of the total mass of these galaxies.

The only alternative to this conclusion is to say that Newton's law of gravity is invalid at galactic distance scales, an option

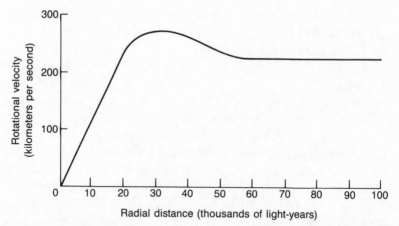

Rotation curve of M31. The rotational velocity does not fall off at large distances beyond the edge of the visible disk. Instead, it levels off at about 200 kilometers per second, indicating that there is a large quantity of invisible matter surrounding the disk.

physicists are loath to take. (Sure, Einstein's theory of general relativity supersedes Newton's law, but it gives essentially the same answer at the relatively low densities encountered in galaxies. The effects of general relativity are markedly different only at extremely high density like that possible in the vicinity of a black hole.) No, there must be "dark halos" of one kind or another hovering about spiral galaxies, unseen shrouds of invisible matter extending well beyond their luminous cores.

For years there had been suspicions about dark halos surrounding galaxies, but it was not until the 1970s that these ideas began to be taken seriously. And it wasn't only the quality of the astronomical observations that helped convince astrophysicists and cosmologists about dark halos; they also satisfied a strong *theoretical* need. At about the same time as Rubin and her colleagues were painstakingly measuring flat rotation curves, James Peebles and Jeremiah Ostriker at Princeton were calculating the gravitational stability of spiral galaxies. They concluded that the disks of these galaxies would fragment unless they were embedded in a roughly spherical halo of unseen matter at least as massive as the visible disk itself.

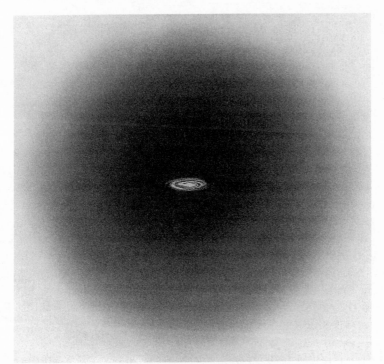

Artist's conception of a spiral galaxy surrounded by a dark halo. Without such a halo around it, the spiral disk would fly apart due to tidal forces.

A thin, rotating disk held together only by its own gravity is unstable, all by itself. Small lumps in its composition — as caused, for example, by groups of stars — would induce vibrations of the disk that would grow steadily larger until it flew apart like a shattered flywheel. If the disk is not alone but embedded in a larger mass surrounding it, however, these disruptive vibrations are damped out and the galaxy is indeed stable.

Thus, the very *existence* of our own Milky Way and similar spiral galaxies requires that there be plenty of dark matter surrounding them. No longer do the observations of dark halos fall on deaf ears. As Sir Arthur Eddington once remarked, "In astronomy, observations are not to be believed until they are confirmed by theory."

——————————— •●• ———————————

As we peer out to still greater distances, we discover that even more of the apparent mass seems not to be there, at least not visibly there. This is the classic "missing-mass" problem astronomers began to recognize as far back as the 1930s. In 1933, Fritz Zwicky, a Swiss astronomer then working at CalTech, noticed that galaxies interacted with one another in ways that suggested the presence of more mass than was visible. Although the phenomenon was known for a long while as the missing-mass problem, this was an unfortunate misnomer. It's not the mass that is missing (if we believe Newton), but the light! The mass *must* be there to make the galaxies move as they do; it is simply not shining.

Take another look at the nearby M31 galaxy, for example, and consider its motion relative to our own Milky Way. The two are the principal members of a small cluster of galaxies commonly known as the Local Group, which also includes several dwarf galaxies. It's a pretty insignificant cluster, as clusters go, but it's the easiest to study in detail. M31 is approaching the Milky Way at a speed of about 100 kilometers per second, or 200 thousand miles per hour. (A race car traveling at 200 miles per hour is going about 100 meters per second, so M31 is going about a thousand times faster.) This is about *half* the speed of the Sun as it travels around the Milky Way. It's an extremely high speed, however, if you consider that the two galaxies are about 2 million light-years apart and assume that this motion is caused by their gravitational attraction for one another. M31 is almost a hundred times farther away than the Sun from the center of the Milky Way, but its relative speed drops by only a factor of *2*.

This is curious indeed. As with planets orbiting the Sun and stars moving about galaxies, Newton's law tells us that relative

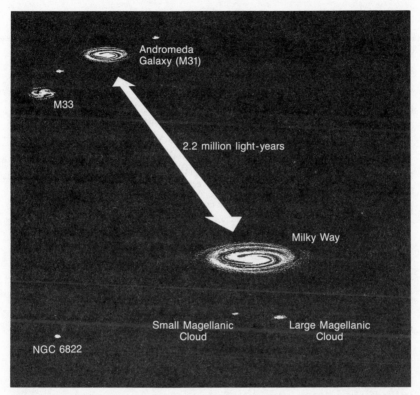

The Local Group of galaxies. The Milky Way and M31 are the two principal members of the Local Group; orbiting them are a total of at least 20 dwarf galaxies, such as the Large (LMC) and Small Magellanic Clouds (SMC).

speeds must fall gradually with distance (if the mass is concentrated at the center). But the flat rotation curves seen at galactic distances now seem to continue far beyond — out to *millions* of light-years. That means there must be extra mass out there, too.

A similar behavior is observed for the globular clusters and dwarf galaxies lying between the Milky Way and M31. Globular clusters are tightly packed spheres of about 100 thousand very old stars orbiting the core of a spiral galaxy out of the plane of its disk. Dwarf galaxies are irregular swarms of hundreds of millions of stars; the Large Magellanic Cloud visible from the

Earth's southern hemisphere is one such dwarf galaxy orbiting the Milky Way at a distance of about 170 thousand light-years. Both globular clusters and dwarf galaxies travel at intermediate speeds of 100 to 200 kilometers per second. They do not add much matter to the intervening space between the two major galaxies of the Local Group, but serve as valuable "probes" of the gravitational field — and hence the mass — that exists there.

What can be going on here? The only way to believe our eyes, within the framework of Newtonian mechanics, is to have unseen mass lurking in the huge "voids" between the two galaxies. Without this extra matter, the relative speeds should all be about one-tenth of what is observed. To make the speeds work out consistent with our observations, there must be about *10 times more* dark matter lying beyond the luminous edge of the Milky Way but still within the orbit of M31. It's as if there were another ten galaxies, each with the mass of the Milky Way, distributed throughout the intervening space.

The same factor of 10 pops when we study the motions of other pairs and small clusters of galaxies. These giant blobs and pinwheels are whirling about one another as if they possessed 10 times the mass that is visible in their shining cores. That means they each carry about a *trillion* solar masses, not 100 billion. The extra 900 billion solar masses, however, are completely dark.

We are forced to conclude that all galaxies are embedded in vast, unseen halos of invisible matter hovering about a core of shining stars packed tightly at their centers. Ninety percent of all the matter in galaxies lurks in these enormous shrouds, which extend out many times the radii of the tiny whirlpools of light located at their pivots.

Now remember that by adding up all the visible mass in the Universe, we obtained a value of 0.005 for the key parameter Ω. If we include the mass of these dark halos in our estimates, we increase Ω by a factor of 10; it is equal to 0.05 instead of 0.005. Given all the uncertainties in these estimates, this value could swell to 0.10 at most: 5 to 10 percent, that is, of the critical density needed to close the Universe. Things look a bit better for a Big Crunch, but we are still far from achieving it. Unless we can find a lot more matter out there in the "emptiness" of space, the Universe is still an open universe.

———————————— • ● • ————————————

What could all this dark matter be? As much as 90 percent of the mass associated with a normal spiral galaxy like the Milky Way seems to be completely invisible. Most of this mass is located in an enormous halo that extends well beyond the luminous disk. Are there configurations of ordinary matter — composed of electrons, protons and neutrons — that can exist without giving off much light?

Black holes, for one, provide an excellent place to secrete a lot of matter without emitting any light at all. Thought to be formed as the end product of stellar evolution when a large star has finally exhausted its nuclear fuel and its core collapses in a tremendous, blinding explosion called a "supernova," a black hole could easily contain many solar masses without ever disgorging a single photon.

One might naively expect black holes to be located in the visible regions of a galaxy, near the other stars, because they are just big stars that finally blinked out. But we can also imagine scenarios in which the very massive stars that end up as black holes are distributed throughout a huge volume as large as the halo, while the smaller stars still shining today are concentrated in its center. There are plenty of reasons why the population of stars formed in the halo might be different from those in the disk.

Another place to hide this dark matter would be in "brown dwarfs," star-like objects with less than 10 percent of the Sun's mass. Intermediate in size between planets and stars, brown dwarfs would not get hot enough in their cores to undergo much thermonuclear burning. Thus, they emit very little light and would be virtually impossible to see unless they are located nearby. Recently reported observations of several possible brown dwarfs (by virtue of their infrared radiation) indicate they may be fairly numerous in the Milky Way.

In the same category as brown dwarfs are planetary blobs the size of Jupiter (with 0.1 percent the mass of the Sun) or smaller, whose cores do not heat up sufficiently to initiate any nuclear reactions at all. To make any sizable contribution to the total galactic mass, however, there would have to be vast numbers of them — far more than the number of stars in the galaxy. Both brown dwarfs and Jupiter-sized planets might well populate the dark galactic halos as well as the visible regions. They could contribute mass to a galaxy, but very little light.

Clouds of gas or dust are other possible forms of halo dark matter—as long as they do not intercept the light observed reaching the Earth from more distant sources. Otherwise, we would realize these clouds were indeed there, from the shadows they made, and add their contribution to the total visible matter —just as we do for the known gas and dust in the disk. Such clouds would have to be arranged in special ways to avoid intercepting this light. But that turns out not to be very difficult, even for very large clouds.

———————————————— • ● • ————————————————

Now let's take a look at some truly enormous celestial objects, the "superclusters," each containing thousands of galaxies swarming about one another. First recognized in the 1950s by the French astronomer Gerard de Vaucouleurs, superclusters are among the largest known structures in the Universe, stretching *many millions* of light-years across space. The Local Group of galaxies belongs to such an ensemble, called the Virgo cluster because its center lies in the general direction of the constellation Virgo. The Milky Way sits near the outer edge of this enormous supercluster, only one galaxy among thousands.

The brightest galaxies at the center of the Virgo cluster. The two large, spherical blobs are massive elliptical galaxies, which are more likely to occur at the cores of large clusters.

The Local Group seems to be falling toward the center of the Virgo cluster, about 60 million light-years away, at a speed of several hundred kilometers per second. Recent measurements indicate this speed may be as high as 600 kilometers per second —or over 1 million miles per hour! From calculations based on Newtonian mechanics, we discover that the mass needed by this supercluster to induce such a high speed corresponds to *several trillion* solar masses per galaxy instead of just the 1 trillion we had estimated from motions within small clusters. And even that quantity was already 10 times the visible mass of a galaxy, so this amount must be several times 10. Thus, the luminous portion of the Virgo cluster is but a paltry few percent of what must be lurking beyond our vision.

Even though there are substantial uncertainties for large systems like superclusters, the general trend still holds true: as we examine larger and larger structures, we require higher and higher amounts of unseen mass to make the motions observed agree with Newtonian mechanics. At these distance scales — of the largest discernible systems — the value of Ω is roughly 0.10 to 0.30, or 10 to 30 percent of what is needed to close the Universe. Although we continue to approach the critical density, we still fall substantially short. (The table below indicates the apparent masses that have been inferred to exist at various distance scales, and the values of Ω associated with each scale.)

Here we have to voice an important caveat. Gigantic systems like the Virgo cluster are not all that common in the Universe. Most galaxies are found gathered in small clusters rather than in such large superclusters. Some cosmologists contend that perhaps these rare superclusters naturally capture more dark matter than do the small clusters. Thus, the mass of an average galaxy may seem a lot larger amidst one of these huge crowds than if it had instead joined a much smaller group. If so, it might be very misleading if we took these giant superclusters to be typical of the Universe as a whole.

APPARENT MASSES AT VARIOUS DISTANCE SCALES AND THE VALUES OF Ω

Celestial object	Typical size (light-years*)	Typical mass (solar masses*)	Apparent value of Ω
Visible part of galaxies	3×10^4	10^{11}	0.005 – 0.010
Dark halos of galaxies	3×10^5	10^{12}	0.05 – 0.10
Clusters of galaxies	10^7	3×10^{15}	0.1 – 0.3
Observable Universe	10^{10}	3×10^{22}	0.4 – 1.6

*1 light-year = 10^{13} kilometers; 1 solar mass = 2×10^{30} kilograms.

If binary galaxies and small clusters are taken to be the most representative systems, then Ω falls back to a value of about 0.05 — and is definitely no more than 0.10. At most, 10 percent of the total mass needed to reverse the expansion of the Universe can be out there, that is, if dark matter clumps preferentially at the scale of a small cluster.

Such a result is consistent with several other ways to estimate Ω, all of which point to a value less than 0.10. This fact was first noticed in 1974 by Richard Gott and James Gunn, then at Cal Tech, and Beatrice Tinsley and David Schramm, then at the University of Texas. In particular, these astrophysicists argued (as we will in Chapter 4) that the relative amounts of hydrogen, helium, and deuterium (heavy hydrogen) observed in the Universe strongly favored a value of Ω between 0.05 and 0.15. "Loopholes in this reasoning may exist," they noted in a caveat, "but if so, they are primordial and invisible, or perhaps just black."

If Ω is indeed about 0.10, the dark matter could simply be ordinary matter made of protons, neutrons, and electrons — which for some curious reason just refuses to glow. (There are plenty of ways normal matter can exist without giving off light, such as the black holes and brown dwarfs just discussed). We need not invoke any fanciful new particles or energy fields to explain such a value.

But there is another possibility, too. When we look at bigger celestial objects like superclusters, we may be examining larger portions of a vast, unseen ocean of dark matter. The motions within binary systems or small clusters can only tell us how much mass there is *inside* their orbits; they say nothing about what lies beyond. Such wider information can be gained only by observing the behavior of bigger systems like superclusters, which is exactly what we did when we moved from the Local Group containing the Milky Way and M31 to the giant Virgo cluster, which includes them both and many other galaxies besides. Perhaps galaxies have enormous halos that extend even farther than we first imagined. They may extend so far, in fact, that we can only see *part* of their gravitational effects even at the scale of a supercluster, where the individual halos of all the galaxies within it have begun to merge into a vast sea of dark matter.

If this second interpretation is true, then the big superclusters may be giving us a more realistic measure of the total dark matter associated with each visible galaxy. In that case they would also be providing a better estimate of Ω. The density of all matter, both visible and dark, would be more like 10 to 30 percent of the critical value—not the 5 to 10 percent we estimated using the motions of galaxies within small clusters. The larger values of Ω thus obtained are difficult (some would say impossible) to explain using only normal atoms. We may therefore have to call upon some kind of dark, exotic particle or relic energy field to make up the difference.

Even at the scale of superclusters, however, we have not achieved an Ω equal to 1. We remain well short of the critical density and still seem to inhabit a Universe headed inexorably for the Big Chill. But studies at larger distances, in general, seem to yield more dark matter and hence larger values of Ω. If we were to continue applying this kind of argument even at distances *beyond* the scale of a supercluster, we might eventually reach the critical density.

There are good indications of structure even at these enormous distances, beyond the 100 million light-year scale of giant superclusters. Astronomers have recently identified long filaments or sheets of galaxies and even a few gigantic voids without a single visible galaxy anywhere within them for hundreds of millions of light-years. Quite possibly, there are some exotic forms of dark matter spread erratically throughout the Universe. Or perhaps even "ghost galaxies"—gigantic clouds of hydrogen gas that somehow failed to condense and form luminous stars. But these interesting possibilities are the subject of a later chapter.

· ● ·

In the last analysis, the motions of many celestial objects — stars, galaxies, clusters, and superclusters — leave no doubt that there is plenty of dark matter in the Universe. To reach this conclusion, astrophysicists have relied on astronomical observations and Newton's laws of gravity and motion. To understand the actual nature of this dark matter, however, will require other arguments that draw heavily upon advances in cosmology and nuclear and particle physics — some of them very recent and speculative. We turn to these arguments in the following chapters.

The single inescapable fact is that all galaxies *do* have dark halos about them — and the bigger ensembles seem to have a greater fraction of their matter that is dark. The actual composition of this dark matter is still a profound mystery that excites the imaginations of astrophysicists and cosmologists today.

4

Origins of the Elements

All about us we perceive a world of splendid diversity, of seemingly endless variety. But the numberless things we can see, hear, feel, and smell are in fact composed of barely a hundred different elements — hydrogen, carbon, nitrogen, oxygen, and iron, to name just a few. These elements are combined with one another in an infinity of ways to produce the apparent complexity of everyday objects.

In the standard Big-Bang cosmology, none of these elements except hydrogen existed at the very beginning. They were all *synthesized*, during the primeval fireball or later, by processes that involved nuclear reactions. How did this synthesis occur? And what can the cosmic abundances of the elements tell us about the earliest moments of creation? These are the questions we turn to now.

By about 1 second into the Big Bang, the Universe had expanded and cooled to the point where nuclear physics could truly begin. The individual protons and neutrons in the primordial soup started sticking together to make heavier, more complex nuclei. Before that moment it was just too hot, with a temperature that exceeded 10 billion degrees, a million times hotter than the surface of the Sun today. Such a high temperature corresponds to an average particle energy of 1 million electron volts (1 MeV) — about a hundred times the energy carried by electrons that smash into a television screen, making the visible images we see there. With such high energies, individual protons and neutrons had been speeding about far too violently to stick together long enough to form heavier nuclei.

But in a short period called the "era of nucleosynthesis," which began at about 1 second after creation and ended about 100 seconds later, our Universe became a tremendous thermonuclear reactor where nuclei of the lightest elements could and did form. It resembled an enormous hydrogen bomb. Almost all our present helium and deuterium, and some of the lithium, were created during that brief stretch of time.

The remnants of these light elements that are detectable today give us an important relic of those early conditions. They prove that our Universe indeed began in a very hot, condensed state. They can also be used to make our best estimate of the average density of normal matter *today*. These light elements therefore provide a sensitive barometer to help us gauge both the past and the future of the Universe.

———————————————— • ● • ————————————————

Just prior to the era of nucleosynthesis, the Universe was a hot soup of electrons and positrons, plus comparable numbers of photons and neutrinos. Sprinkled throughout this vast cauldron like salt and pepper in a piping hot soup was a light seasoning of protons and neutrons. For every proton or neutron, there were at least a *billion* photons dashing about—and perhaps as many as 10 billion. All the heavier, highly unstable baryons had vanished. They survived much less than a microsecond (or a millionth of a second), and there was simply not enough energy left to continue replacing them as fast as they disintegrated.

The composition of the Universe during this epoch, from a hundredth of a second to about 100 seconds after creation, was not too different from the conditions studied here on Earth by nuclear physicists, so its evolution is well understood. How matter behaves under similar conditions has been examined repeatedly in nuclear research laboratories around the world. There is nothing mysterious about what occurs. We have the greatest confidence that we can make valid statements about the cosmology of this epoch based on our current knowledge of nuclear physics.

A key to understanding the physics of this period is the concept of thermal equilibrium. This phenomenon represents a delicate state of balance between opposing reactions, be they chemical or nuclear. Whenever a system is in thermal equilibrium, its *temperature* alone determines the relative quantities of the different interacting species—subatomic particles, nuclei, atoms, or molecules—that are present. (Temperature is just a gross measure of the average "kinetic energy," or energy of motion, of a collection of particles. A temperature of 10 billion degrees, for example, means that the particles are moving with an average kinetic energy of about 1 MeV.) The Universe was in thermal equilibrium just before the era of nucleosynthesis, and therefore its temperature governed the ratio of neutrons to protons, which we write as n/p.

When the temperature was *above* 10 billion degrees, which was the case before the Universe was 1 second old, the numbers of neutrons and protons were roughly equal, because these particles were easily converted into each other. A neutrino striking a neutron produced a proton plus an electron about as easily as the opposite reaction—an electron plus a proton making a neutron plus a neutrino. And an antineutrino (the antiparticle of a neutrino) striking a proton gave back a neutron plus a positron about as often as the reverse reaction. As long as the early Universe was hot enough so that neutron-producing and proton-producing reactions balanced, their numbers remained equal, or $n/p = 1$.

This cozy state of affairs lasted until the Universe was about 1 second old. By that time the temperature had fallen to 10 billion degrees, and the reactions producing neutrons from protons were slowing down. The Universe still remained in equilibrium, but the balance between the neutron-producing and proton-producing reactions was shifting. A neutron is a bit *heavier* than a proton—about a tenth of a percent, or 1.3 MeV in energy units (here we are using the equivalence of mass and energy implied by Einstein's famous formula, $E = mc^2$). When an electron and proton collide, therefore, they must supply the additional energy needed to make up this mass difference; otherwise

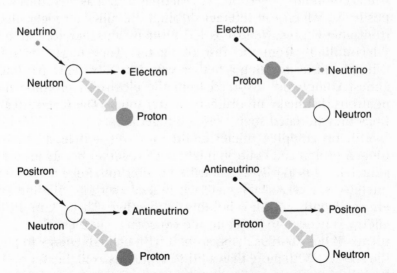

Nuclear reactions that occurred at about 1 second into the Big Bang. At this moment the reactions producing neutrons from protons were just beginning to slow down relative to the proton-producing reactions.

a neutron cannot form. Without that extra input they just rebound harmlessly away. At about 1 second after creation, when the average particle energy was around 1 MeV, the ratio n/p of neutrons to protons had fallen to $1/3$. Neutrons, that is, made up around 25 percent of the baryons in the Universe, and protons contributed the remaining 75 percent.

At about the same instant, the weak nuclear force driving all these reactions was losing its effectiveness. Recall from Chapter 2 that this force has an appreciable impact only when subatomic particles are very energetic and come extremely close together. As the headlong explosion of matter continued beyond the first second, however, these conditions were no longer met. The various subatomic particles were rapidly losing energy, and their density had fallen too far for the weak force to have much effect upon them.

Cosmologists say that neutrinos "froze out" at this moment, about 1 second after creation. Because they can feel only the effects of the weak force and none other, neutrinos and antineutrinos could no longer initiate reactions as fast as electrons and positrons, which can interact through the much stronger electromagnetic force. Neutrinos and antineutrinos therefore started "decoupling" from the rest of matter. They began drifting through space, forming a nether world of subatomic particles almost completely divorced from the electrons, protons, and neutrons that make up ordinary matter today. The reverse reactions they initiated soon ceased altogether.

Still, no complex nuclei could form yet — at least not for long. A proton and neutron might stick together briefly to make a nucleus of heavy hydrogen called deuterium. But a deuterium nucleus is a very shaky marriage indeed, with a "binding energy" of only 2.2 MeV holding it together. That's how little energy is needed to break up the twosome, and there were still plenty of hot photons flying around with enough energy to play the spoiler. With more than a billion photons available for every proton or neutron, *some* photons always had more energy than the 2.2 MeV needed.

As the Universe kept expanding and cooling, the ratio of neutrons to protons continued to drop. An individual neutron can disintegrate into a proton, an electron, and an antineutrino, increasing the supply of protons and decreasing the number of neutrons left. Protons, however, do not decay; as far as we know, they live forever. So after about 100 seconds, there was only one neutron left for every seven protons, or $n/p = 1/7$.

Meanwhile the temperature of this particle soup had cooled to about 1 billion degrees by that time, corresponding to an average particle energy of 0.1 MeV. While the baryon population was busy converting from neutrons into protons, the population of positrons was dying off and the number of electrons was becoming comparable to that of protons. These changes were occurring because it had become difficult to recreate positrons as fast as they disappeared. When an electron and positron meet, they annihilate one another — leaving behind only pure energy in the form of photons. A photon can regenerate an electron – positron pair when it smashes into a baryon, thus replenishing the positron supply, but it must have an energy of at least 1 MeV (and preferably more) to do so. Photons energetic enough to accomplish this were then becoming increasingly scarce, although there were plenty of hot miscreants left to wreck any fragile neutron – proton fusions. Since there were over a billion photons for every baryon, even a tiny fraction of energetic photons was enough for the task.

Suddenly, at an age of about 100 seconds, the temperature of the Universe had dropped to the point where a proton and neutron *could* stick together to form a deuterium nucleus and not be immediately torn asunder. In a split second, the available neutrons were all swept up in a great mating frenzy, which did not stop with twosomes. More stable unions can be formed with three or four members — two neutrons plus a proton to make a nucleus of tritium, two protons plus a neutron to make helium-3, or two protons and two neutrons to make helium-4. So just as quickly as deuterium could form, it became absorbed into tritium and two forms of helium.

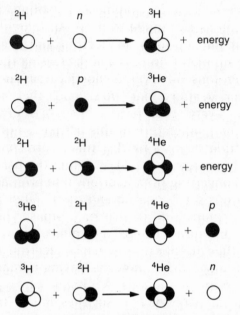

Nuclear reactions that led to the formation of tritium (^{3}H), helium-3 (^{3}He), and helium-4 (^{4}He) during the era of nucleosynthesis — at about 100 seconds into the Big Bang. Almost all the protons and neutrons then present ended up as ordinary hydrogen (p) and helium-4, with traces of deuterium (^{2}H), helium-3, and lithium remaining.

The most stable unions were those of helium-4, or ^{4}He in physicists' shorthand, nuclei with two protons and two neutrons apiece. With a binding energy of 28 MeV (or 7 MeV per baryon), they are virtually unbreakable at 1 billion degrees. So moments after the so-called "deuterium bottleneck" had been breached, almost *all* the available neutrons were locked up in these close-knit foursomes. (Deuterium is not special in any way. It's just the first nucleus to form, when the temperature becomes low enough for nucleosynthesis, because it only requires one proton and one neutron to meet.) Combinations of five or eight members disintegrate immediately, making it virtually impossible to form more complex nuclei, so the fusion process stopped almost as abruptly as it had begun.

As the era of nucleosynthesis ended, essentially all the baryons in the Universe existed either freely as single protons or were trapped inside of helium-4 nuclei. Some tiny residues of deuterium (^{2}H), tritium (^{3}H), and helium-3 (^{3}He) remained, along with a scant trace of lithium-7 (^{7}Li), which has three protons and four neutrons per nucleus.

The fraction of helium-4 nuclei in the Universe depends on the ratio n/p of neutrons to protons in existence at the exact moment (shortly after it was 100 seconds old) that the deuterium bottleneck was finally breached and complex nuclei could begin to form. If there was one neutron for every seven protons at that instant, then we expect that one-quarter of these baryons were swept up into helium-4, because each neutron takes a proton with it into bondage, leaving the other six to roam about fancy-free. Two out of every eight baryons in existence, that is, were locked up inside helium-4 nuclei.

Because protons and neutrons have essentially the same mass (and are far heavier than electrons), about two-eighths (one-quarter) of the normal matter in the Universe ended up as helium. One-hundred thousand years later, when things had cooled to the point where the remaining electrons could finally bind to these primordial nuclei and form atoms, the helium mass fraction did not change noticeably. Electrons have less than a thousandth the mass of a baryon, so adding one or two of them made little difference.

———————————————— • ● • ————————————————

Only the very lightest elements were produced during the Big Bang. All the heavier elements — the carbon, nitrogen, and oxygen in our bodies and in the air we breathe; the silicon, aluminum, copper, and iron in common appliances and automobiles — were forged afterwards by hot thermonuclear fires burning in stellar ovens. Ordinary stars like our Sun cook hydrogen to make helium. Others, having exhausted their hydrogen, burn helium

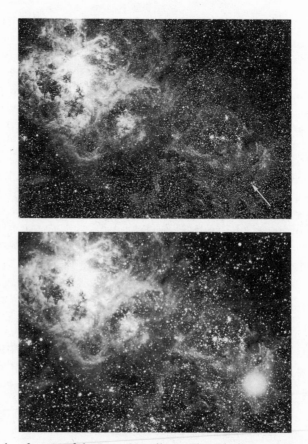

Photographs of a part of the Large Magellanic Cloud before and after February 23, 1987. The supernova SN1987A is clearly visible as the bright spot at the lower right of the bottom photo. The arrow in the top photo indicates the star that became the supernova.

to make carbon, oxygen, and a host of heavier elements. Eventually these elements are spewed into space by giant stellar explosions called supernovae, like the one dubbed SN1987A that appeared in the Large Magellanic Cloud on February 23, 1987. These processes have been occurring ever since stars first formed shortly after the initial fireball cooled. Yet all the elements heavier than helium make up less than 2 percent of the visible matter in the Universe. The other 98 percent is the original, primordial hydrogen and helium left over after the era of nucleosynthesis ended.

Astrophysicists employ a variety of techniques to determine the relative abundance of the different elements in the Universe. They examine meteorites or moon rocks, for example, to tell how much of the heavy elements like iron were present at the origin of the Solar System about 4.5 billion years ago. But gaseous elements like hydrogen or helium are more elusive. Because they do not remain trapped in meteorites for long, we are forced to use other means to establish their relative abundances.

The best way to examine hydrogen and helium is to study the visible and ultraviolet light emitted by stars in our own and in nearby galaxies. When atoms of a particular element are heated, they emit electromagnetic radiation at a few specific wavelengths, or colors, that are characteristic of that element. (Each wavelength corresponds to a difference between two possible energy levels of the atom, as discussed in Chapter 2.) Astronomers study this starlight with a "spectroscope," which employs a prism to spread out the spectrum of colors contained in the light. The characteristic colors of hydrogen and helium atoms show up as bright lines — "spectral lines" — appearing in the spectrum. By comparing the intensities of the helium lines with those of the hydrogen lines, one can establish the *relative* amounts of these two elements present in the star.

Such a measurement, of course, may not tell us much about their relative abundance in the Universe as a whole. If we look at a star that has already cooked much of its hydrogen into helium, for example, we might overestimate the overall helium abundance. We have to look at younger stars, and in areas of space that have not yet evolved very much, to get a better idea of the primordial abundances. (For the spectral lines of helium to be observable, we require a very hot gas, which means we must look at very hot celestial objects like young, massive stars.) Or we can look at what happens to starlight as it passes through interstellar gas. In such a case the hydrogen and helium present *absorb* light at the same characteristic wavelengths, leaving dark "absorption lines" in the spectrum observed in a spectroscope. Again, the relative darkness of these lines — those of hydrogen versus helium — tells us the ratio of the two elements present in the gas.

By the early 1960s such spectroscopic measurements were becoming increasingly consistent, revealing that about 25 percent of the visible matter in the Universe was helium-4, and that almost all the remainder was hydrogen. Of course, stars manufacture helium from hydrogen, so scientists still had to be careful interpreting this number. But detailed studies of the heavier elements detectable indicated that the vast majority of the helium could not have been produced in stars, which make helium along with other elements. Because the total of all these heavier elements is less than 2 percent of the visible mass of the Universe, not much stellar nucleosynthesis can have occurred yet. In other words, if much of the helium had been produced in stars, we would see large amounts of heavy elements, too, and we do not. Therefore almost all the helium observed had to be *primordial* helium.

During the late 1940s George Gamow, Ralph Alpher, and Robert Herman had suggested that helium would have been produced in the Big Bang. But it was not until the mid-1960s that Fred Hoyle and Roger Tayler at Cambridge, and later James Peebles and CalTech's William Fowler and Robert Wagoner, were able to make accurate calculations of primordial nucleosynthesis. Their successful explanation of the helium abundance was a great triumph for Big-Bang advocates. Along with Penzias and Wilson's 1964 observation of the 3°K background radiation, these arguments helped convince many scientists that the Universe had emerged from a very hot, violent explosion. Indeed, Gamow and his two colleagues had used their own nucleosynthesis arguments to estimate that the present temperature of the Universe should be a few degrees, years before Penzias and Wilson measured it in 1964.

The steady-state theory originally championed by Hoyle, in which matter was supposedly being created continuously, soon fell by the wayside. Ironically, the leading advocate of this rival cosmology played a key role in its eventual demise.

–––––––––––––––––––––––––––– • ● • ––––––––––––––––––

Astrophysicists did not stop with calculating the helium-4 abundance due to Big-Bang nucleosynthesis. In the early 1970s, they also realized that leftover traces of primordial deuterium and helium-3 provided a sensitive means of determining the actual *density* of the Universe during the era of nucleosynthesis. By a simple extrapolation forward in time, assuming a uniform Hubble expansion, they could obtain the average density of matter in the Universe today — and therefore predict whether it would continue to expand forever or eventually collapse.

During the 1960s, astrophysicists had generally believed that all light elements except helium-4 were synthesized in stars. This belief was based on a scenario originally developed by Fowler and Hoyle, who claimed that any deuterium and lithium observed had been produced during star formation. Fowler — a jolly, often bearded, roly-poly physicist easily mistaken for the neighborhood bartender — had previously shown how heavier elements were produced in stars, a feat for which he shared the 1983 Nobel prize. At least initially, nobody questioned his scenario for making deuterium and lithium.

Then in the early 1970s, a group of astrophysicists in Paris led by Hubert Reeves proved that newborn stars do not have enough energy (per baryon) to produce the nuclei of these light elements. With the countenance of a gnome and a highly persuasive speaking manner, the Canadian-born and Cornell-educated Reeves has established himself as a leading scientific spokesman on French television — their equivalent of Carl Sagan. In the summer of 1970, Reeves teamed with Fowler and Hoyle to suggest a different source of the light nuclei: cosmic rays. But while they could thereby explain *some* of the lithium (plus other light nuclei like beryllium and boron), their proposal failed completely when it came to deuterium.

At about this time an important experiment performed during the Apollo 11 Moon landing was being analyzed by Johanas Geiss in Switzerland. In this test the astronaut Neil Armstrong had unfurled a metal foil at the Moon's surface and taken a sample of the solar wind — a steady stream of protons and other subatomic particles spewing out in all directions from the Sun. These particles adhered to the foil, which Armstrong rolled up and brought back to Earth on his return. Geiss' analysis of this foil revealed that the Sun contained *one-tenth* as much deute-

Astronaut Neil Armstrong deploying a metal foil on the Moon's surface during the Apollo 11 lunar landing in 1969. Particles in the solar wind adhered to the foil, which was later analyzed by Johanas Geiss to determine the abundance of deuterium in the Sun.

rium as seawater, which had been used previously to estimate its cosmological abundance. And because earthbound water would be biased toward a higher deuterium concentration (due to its relative ease of chemical formation), Geiss' measurement gave the better estimate. Thus, the cosmological abundance of deuterium suddenly plummeted by a factor of 10.

Jean Audouze, who had been a student of Reeves, came to Cal Tech as a postdoc in 1971, bringing with him the surprising news from Switzerland. He and David Schramm, then a postdoc working with Fowler, soon recognized that such a tiny amount of deuterium could easily have been produced in the Big Bang, whereas the large quantity based on the old seawater estimate would have been difficult to generate. There was no longer any pressing need to invent other sources of deuterium, such as newborn stars or cosmic rays.

Whether or not the Big Bang was the only possible source was finally resolved a few years later when Schramm and his students showed that deuterium nuclei can only be *destroyed* in stars, not created. And the solar-wind measurements of the deuterium abundance were confirmed in 1973 by the spectroscopic analysis of ultraviolet starlight using the Copernicus satellite.

So all the deuterium detectable today must be primordial, a key relic of the Big Bang. And the denser the Universe happened to be during the era of nucleosynthesis, the less deuterium would have survived, because more of it would have ended up as helium. (At higher densities the deuterium nuclei can "find" one another more readily and then stick together to form helium-4). Using the observed present-day abundance of deuterium, which is about 20 parts per million, astrophysicists therefore established an upper limit on the density of matter in the Universe today. It must be smaller than 5×10^{-31} grams per cubic centimeter. Any greater, and there would be less deuterium remaining today than is actually found.

By contrast with deuterium, nuclei of helium-3 are *created* in stars, not destroyed. Therefore, its present abundance in the interstellar medium — also around 20 parts per million — represents at least the amount of primordial helium-3 that was created in the Big Bang. Stellar processes can only have increased the total. This observation allowed a group of astrophysicists to establish a *lower* limit on the density of matter in the Universe. It must be greater than 2×10^{-31} grams per cubic centimeter. Any lower, and there would be more helium-3 around than we witness today.

What was truly remarkable about these two independent limits — based on the abundances of deuterium and helium-3

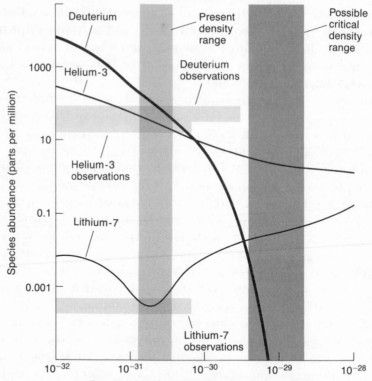

Comparisons of observed abundances (lightly shaded areas) of deuterium, helium-3, and lithium-7 in the Universe with nucleosynthesis calculations. The calculated abundances (curved lines) depend strongly on the present density of nucleons. Agreement is achieved only for the narrow range between 2 and 5×10^{-31} grams per cubic centimeter, which is still about a factor of 10 to 20 below the critical density.

—was the fact that, taken together, they permitted only a narrow range of possible densities. These limits tightly constrained the density of the Universe during the era of nucleosynthesis. So by straightforward extrapolation, we can conclude that the average density of matter in the Universe today is no more than 10 percent of the critical density.

These density limits have been gradually refined over the past decade, getting tighter and more convincing almost every year. During the early 1980s, measurements of the lithium-7 abundance in old stars (less than a part per billion) became accurate enough to confirm the conclusions reached earlier using deuterium and helium-3. The amount of lithium-7 that could have been produced in the Big Bang was consistent with these observations if and only if the density of the Universe fell into the *same* narrow range as the one discussed above. Calculations based on Big-Bang nucleosynthesis can thus reproduce the concentrations of the light elements — ranging over ten orders of magnitude — extremely well. This concurrence between theory and measurement is a strong validation of such arguments.

These density figures first emerged during the mid-1970s, when a number of other observations pointed to an open, infinite universe. There seemed to be far too little matter in the Universe, one-tenth as much as needed, to ever halt its relentless Hubble expansion. Matter would continue fleeing matter forever, or so it seemed. Space apparently stretched out to infinite distances without any boundaries. "In all dimensions alike," as the Roman poet Lucretius had noted 20 centuries earlier, "on this side or that, upward or downward through the Universe, there is no end."

———————————— • ● • ————————————

What astrophysicists and cosmologists had unwittingly overlooked was the possibility of *other* forms of matter quite unlike the ordinary stuff we encounter in everyday life. We have alluded to this possibility in earlier chapters, and now begin to discuss it in earnest. Arguments based on Big-Bang nucleosynthesis are strictly valid only for ordinary, baryonic matter, whose mass is due predominantly to the familiar neutrons and protons that participate in nuclear reactions. They say absolutely nothing about any "nonbaryonic" matter — other, stranger kinds of particles that do not participate in these reactions at all.

Elementary-particle physicists of the late 1970s were beginning to propose a number of new possibilities. Neutrinos, for example, the light, neutral cousins of electrons and positrons, had been ignored because almost everyone considered these motes to be absolutely massless. But give them even a tiny mass, less than a ten-thousandth the mass of an electron, and these ghostly particles can easily dominate the total mass of the Universe because there are so unbelievably many of them around. With thousands permeating every cubic inch, or hundreds per cubic centimeter, they are *billions* of times more plentiful than electrons, protons, or neutrons.

All that we can conclude from arguments of Big-Bang nucleosynthesis, in fact, is that the density of *baryonic* matter in the Universe is less than about 10 percent of the critical value. The ratio Ω of the total density of everything to the critical density could still be equal to or even greater than 1 if there were enough additional nonbaryonic matter around. Whether or not the Universe is closed remains an open question.

Although it may be just a coincidence, this same 10 percent figure is roughly consistent with the total amount of mass that apparently collected in galaxies, as described a chapter ago in the discussion of galaxy motions. All the matter in galaxies, including both their visible cores and the halos of dark matter surrounding them, can only contribute about 10 percent of what is needed to close the Universe. As far as we can tell, therefore, the dark halos around individual galaxies might well be normal, baryonic matter that for some reason just does not shine. There is no hard-and-fast reason to require that it be anything more exotic than black holes, brown dwarfs, large planets, or huge clouds of gas or dust.

Primordial black holes, which formed not from the collapse of a huge star but during the first second of the Big Bang, would not be subject to the limits imposed by these nucleosynthesis arguments. The baryonic matter trapped in such an object would have bypassed the deuterium and helium formation that occurred during the era of nucleosynthesis. But primordial black holes have a nasty habit of evaporating due to quantum mechanical processes, as shown by the British physicist Stephen Hawking. Unless a primordial black hole originally had a mass greater than that of a small asteroid, or about a billion tons, it would have vanished into pure radiation by now. So black holes larger than an asteroid (but smaller than the Sun) give us a way to obtain more than 10 percent of the critical density using baryons alone, but the excess is still locked up inside them, unavailable for our inspection.

Many cosmologists have tried to find other loopholes in the Big-Bang nucleosynthesis arguments that would permit a larger baryon density today. In the late 1980s, several scientists suggested that there could have been density fluctuations during the first second after creation, before the era of nucleosynthesis even began. Preliminary calculations showed that the total baryon density might indeed be higher in such a scenario. But more detailed studies (which accounted for interactions between low-density and high-density pockets of the early Universe) proved that the baryon density still had to be less than about 10 percent of the critical value in order to account for the observed abundances of light elements. These calculations showed once again how robust were the conclusions about Big-Bang nucleosynthesis.

On the scale of superclusters stretching almost 100 million light-years across space, the value of Ω deduced from galaxy motions might be larger than 0.1 — perhaps as large as 0.3 or more. If it happens to be much more than 0.1, the total mass cannot be due to baryons alone. We would need some kind of nonbaryonic matter to make up the difference. Baryonic matter may be enough to account for the mass that appears to hover around individual galaxies, yielding $\Omega = 0.1$ at most. But it will not suffice if there is a lot more mass distributed at larger distance scales. Big-Bang nucleosynthesis arguments tell us that there must be something *else* out in intergalactic space if we need to achieve values of 0.2 or 0.3.

————————————— • ● • —————————————

Big-Bang nucleosynthesis is so well understood today that these arguments can be turned around and used to determine how many different kinds of neutrinos there are in nature. Although they froze out of equilibrium — or decoupled from ordinary matter — when the Universe was about 1 second old, neutrinos (and antineutrinos) still had an important role to play in its subsequent evolution. At that instant they were as common as light, and equally energetic. They carried a large fraction of the mass – energy in the Universe. In a subtle relationship discovered in the mid-1970s, the neutrino population had a small but important effect on the abundance of helium-4. As measurements of this abundance became more and more accurate, they were used to determine how many different kinds of neutrinos there are.

The primordial helium-4 abundance depends critically upon the time that elapsed between the moment neutrino freeze-out occurred and the instant atomic nuclei could form and remain intact. Remember that the ratio n/p of neutrons to protons fell steadily during this interval, from a value of about 1/3 to about 1/7, because neutrons were gradually converting into protons. The less time that elapsed between the two key events, the more neutrons there remained at the end to form deuterium and helium nuclei. And the faster the Universe expanded, the shorter this interval would have been, which means that n/p could not have fallen as far. In addition to this effect, a faster expansion rate means that the neutrinos would have frozen out earlier, at a higher temperature, and the ratio n/p would have begun its inexorable decline at a higher value. Both effects would have combined to drive the neutron fraction higher, and more neutrons means more helium-4 would have been produced.

Here is where the neutrinos come in. The expansion rate of the early Universe is closely related to the number of "relativistic" particles — those moving at or near the speed of light — that were present during this brief interval. Neutrinos, being light and speedy, qualify. So do photons, which travel at the speed of light by definition. Electrons and positrons qualify, too, but just barely. The greater the number of different species of relativistic particles (those with masses less than about 1 MeV), the faster the expansion of the early Universe and the less time available for neutrons to convert into protons. So additional kinds of neutrinos means *more* helium-4 would have been produced in the Big Bang.

The primordial helium-4 abundance is, in effect, a barometer of the number of different kinds of light, relativistic particles present at creation. Besides the photons, electrons, and positrons, there were at least the three known types of neutrinos — electron, muon, and tau neutrinos (plus their antiparticles) — that qualified as relativistic particles. Additional species of neutrinos would have induced a slight increase in the helium abundance, as long as they were moving relativistically during the era of nucleosynthesis. So would any other light, exotic beasts not included in the Standard Model, as long as they were traveling at close to the speed of light. But anything heavier than about 10 MeV, whether neutrino or not, would have had no effect at all upon the amount of helium produced. The nucleosynthesis arguments tell us *nothing* about these or any heavier particles.

How well can astrophysicists determine the abundance of primordial helium? This is a key number employed in these arguments. If they take the helium fraction observed today, and subtract the small percentage that must have been forged inside stars, they get a result less than 25 percent. The "best-fit" value is about 23 percent, so the primordial helium abundance is fairly well constrained by observations.

Using this range of values in 1983, the Chicago cosmology group (which included David Schramm), together with Gary Steigman of Ohio State University, determined that there are probably only three different species of light neutrinos. According to their analysis, a fourth kind of neutrino was just barely possible, but not likely. The three kinds of neutrinos already known (from particle physics experiments, that is) are all that are needed to produce the amount of helium observed. (Here we are assuming that the tau neutrino is indeed light, even though the experimental upper limit on its mass is currently 35 MeV. Remember that only particles with masses less than 10 MeV have any effect on the helium abundance.)

Because there can only be a single neutrino in each family of leptons and quarks (see Chapter 2), the nucleosynthesis arguments restricted the total number of elementary particles al-

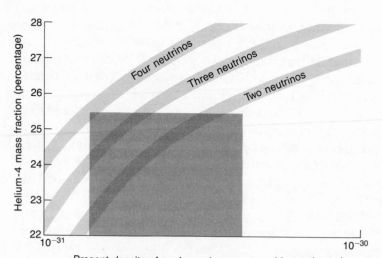

Present density of nucleons (grams per cubic centimeter)

Calculated and observed abundances of helium-4, plotted versus the density of nucleons in the Universe. The calculated abundances depend on the number of different neutrino types assumed to exist. Agreement occurs only if there are two or three types; a fourth was just barely possible when these calculations were made in the mid-1980s, but further refinements of the helium measurements and neutron lifetime now exclude that possibility.

First family	Second family	Third family	Fourth family(?)
• up quark	● charm quark	⬤ top quark	(?)
• down quark	● strange quark	● bottom quark	(?)
• electron neutrino	• muon neutrino	○ tau neutrino	(?)
• electron	● muon	● tau	(?)

The fundamental constituents of matter, as allowed by particle physics and cosmology by the mid-1980s. According to the nucleosynthesis arguments, there could have been at most four different families of quarks and leptons.

lowed in the Standard Model. At most, there could be four families, which would allow four charged leptons and eight quarks in all. If, however, there were just *three* families, then the electron, muon, and tau would be the only charged leptons, and only the top quark remains to be found among the six possible quarks.

When these kinds of arguments were first published, in a 1976 paper Schramm and Steigman wrote with James Gunn, particle physicists hadn't the slightest idea how many fundamental quark–lepton families there were in all. There could have been infinitely many, for all we knew. More than eight, however, would have been a disaster for the dominant theory of the strong force, known as "quantum chromodynamics," or QCD, which explains how gluons behave.

At the time, Gunn, Schramm, and Steigman argued there could be no more than seven, which brought a measure of relief to the particle physics community. The limits were looser then than they are today because of ambiguities in the calculations and uncertainties in the primordial helium abundance. Subsequent improvements in both areas allowed the narrower limits published in 1983. As far as cosmologists were concerned, there could only be three or four families.

———————————— • ● • ————————————

For a long time elementary particle physicists were skeptical of the cosmological limits on the number of fundamental families. They considered the nucleosynthesis arguments to be based on shaky assumptions that might well prove false, thus invalidating the principal conclusions. There may have even been a bit of professional jealousy involved, too. Who did these upstart cosmologists think they were, anyway, making predictions about the number of fundamental building blocks using some nebulous arguments about the Big Bang? Whatever the case, most particle physicists preferred to wait until experiments at accelerators could settle the matter once and for all.

Answers began to come in during 1986, proving that cosmologists were not too far off the mark. By that year CERN's big proton–antiproton collider, in which the Z particle had been discovered in 1983, had produced about a hundred of these massive entities, the heaviest known subatomic particle. By making the first crude measurements of the Z particle's lifetime, a group of physicists led by Carlo Rubbia were able to conclude that there could be no more than five different kinds of neutrinos — and that *three* was the likeliest number.

Among other things, the Z particle breaks up into pairs of neutrinos and antineutrinos. As long as they are lighter than half the Z mass, extra kinds of neutrinos would give the Z particle additional ways by which it could disintegrate, shortening its lifetime slightly. The situation is a bit like a bucket full of water with bullet holes in it; shoot an extra hole in the bucket, and the water drains away a little faster. An accurate measurement of the Z lifetime therefore reveals exactly how many different kinds of neutrinos exist.

Other experiments at electron–positron colliders around the world reached similar conclusions about the number of neutrinos. By examining those collisions where an electron and a positron annihilated one another to make a pair of (undetected) neutrinos plus a high-energy photon, physicists concluded there could be no more than five different kinds of neutrinos in all. If the cosmologists were wrong, they could only be a little wrong.

A decisive answer to this important question had to wait until 1989, however, when copious quantities of Z particles finally became available in the United States and Europe. As the year began, the world's supply of Z's consisted of a few hundred produced in proton–antiproton colliders built at CERN and Fermilab. But by year's end two new electron–positron colliders — dubbed the SLC and LEP — had begun operations at SLAC and CERN, able to generate this massive particle by the thousands.

Led by Burton Richter, SLAC physicists refitted their aging 2-mile linear accelerator to boost bunches of electrons and positrons to almost 50 GeV (50 gigaelectron volts, or 50 billion electron volts), and added a pair of curved tunnels loaded with magnets to guide and focus these particles into collision at the heart of a huge particle detector. To enhance the chances of collision, the SLC had to compress the bunches to about 6 microns across — much less than the width of a human hair. By October 1989, a team of physicists operating the detector had observed the remains of almost 500 Z particles created when an electron and a positron with the right energies annihilated each other. Based on this sample they concluded that there were only *three* different kinds of neutrinos, and thus only three conventional quark–lepton families in the Standard Model.

Hot on the heels of the SLAC team were many hundreds of physicists at CERN, who were building LEP — the world's largest particle collider. A huge storage ring 27 kilometers, or 16

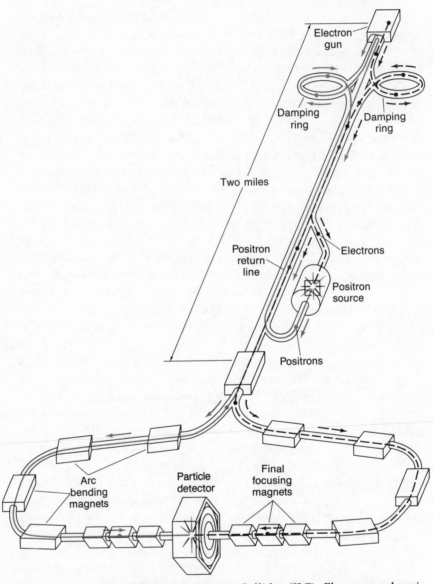

Artist's conception of the Stanford Linear Collider (SLC). Electrons and positrons are boosted to high energy in the 2-mile linear accelerator, then swung through two curving tunnels before colliding to produce Z particles inside a large detector.

miles, in circumference, this colossal machine rests in an enormous tunnel drilled beneath the sleepy farms and villages dotting the broad plain west of Geneva, Switzerland. Compact bunches of electrons and positrons race in opposite directions around the ring, passing through one another thousands of times per second, and creating a Z particle about once a minute. LEP began operations in August 1989 and had amassed 11,000 Z's by October. One day after SLAC announced its conclusions, four different groups of physicists working on LEP confirmed that there were indeed just the three known kinds of neutrinos, and no more. In 1989, also, improved measurements of the helium abundance and neutron lifetime had been used by cosmologists to conclude there were only three light neutrinos, so everything came together nicely that year.

While confirming cosmological predictions made years earlier, these particle physics experiments made much stronger statements. Any small possibility of a fourth neutrino was convincingly ruled out by the SLC and LEP, even if it was as much as *30* times heavier than the proton. The Big-Bang nucleosynthesis arguments hold true only for particles whose mass is less than 10 MeV, or only about 1 percent of the proton's mass. By limiting the different kinds of neutrinos, these experiments also put a firm cap on the complexity of nature — at least at the level of its fundamental building blocks. There can be only three conventional families of quarks and leptons in the Standard Model.

• ● •

So the cosmologists were essentially right, after all. The fact that cosmology could successfully limit the number of elementary entities is a marvelous demonstration of its growing interplay with particle physics. Previously cosmologists had used information from particle physics to make conclusions about the conditions in the early Universe, but now things were working the other way, too. Events that occurred in the first few moments of existence indicated that there could be only so much complexity in the standard particle table. Any more, and the Universe would look different today. This is perhaps the first time since Newton's day that a cosmological argument has provided insight into fundamental physics.

5

The Creation of Matter

When Steven Weinberg wrote his classic book *The First Three Minutes* in 1976, he began his account of the Big Bang at a time of 0.01 second, one-hundredth of a second. At this point the temperature of the primeval fireball had fallen to 100 billion degrees, over ten thousand times hotter than the core of the Sun today. The density of the universe then was absolutely stupendous—about a *billion* times denser than water. A tiny droplet of this seething plasma as small as a grain of sand would have weighed as much as an automobile. The behavior of matter and energy under such extreme conditions was nevertheless well understood by 1976, so Weinberg could describe the evolution of the Universe from that moment on with supreme confidence. As we did in the last chapter, he only needed to call upon some well-established laws of nuclear physics to tell his intriguing tale.

Weinberg was more cautious about what had happened during the course of that first hundredth of a second, however, and he therefore confined his speculations about this earliest period to a short chapter at the end of the book. As time is rolled backward into this brief instant, the temperature climbs to the point where a plethora of heavy mesons and baryons can be created from pure energy. These particles exert the strong force upon one another. This force, which is what binds atomic nuclei together, was still somewhat of a mystery in 1976. "We simply do not yet know enough about the physics of [these] elementary particles to be able to calculate the properties of such a melange with any confidence," lamented Weinberg. "Thus our ignorance of microscopic physics stands as a veil, obscuring our view of the very beginning."

Today we understand more about how elementary particles behave, or at least we have much more confidence that certain theories available in 1976 are indeed *correct*. As mentioned in Chapter 2, particle physicists have proved that mesons and baryons are not elementary particles at all, but are instead aggregates of far tinier entities called quarks. And we have an exact theory, quantum chromodynamics (which is incorporated in the Standard Model), that tells us how these quarks interact with one another when they get very close together — as must have been the case during the earliest moments. The great advances in particle physics over the past two decades permit us to peer behind Weinberg's veil and glimpse the act of creation itself.

Understanding the behavior of quarks, however, has not resolved *all* of the mysteries about that first hundredth of a second. How was matter created in the first place? Why was it so smoothly distributed and uniform in all directions, like an infinitely large mass of whipped cream? How did galaxies begin to coalesce then, if everything was initially so smooth? And where did the dark matter come from? The ultimate answers to these questions must be found by examining the origins of time itself.

Until the late 1970s, the origin of matter proved a major puzzle for Big-Bang cosmology. Although there is only a single baryon or electron for every *billion* or so photons in the Universe, even this tiny amount of matter seemed curious. The laws of physics are symmetric between matter and antimatter, but the Universe itself was apparently not. In a high-energy physics experiment such as those done at particle accelerators, scientists can create matter only if they create an exactly equal amount of antimatter. An energetic photon can be converted into an electron and a positron, for example, and the two can annihilate one another to yield pure energy. But nowhere did it ever seem possible to create a *net* amount of matter over antimatter. So how did the excess matter observed in the Universe today ever arise? Did it exist from the very beginning?

You might question how we *know* for a fact that there is indeed an excess of matter over antimatter. Might there not be separate pockets of each isolated from one another, and we just happen to inhabit a pocket of matter? The answer to this question, which began coming in during the 1970s, is a resounding no.

In 1970 Gary Steigman, working at Hoyle's Cambridge institute, and Yacov Zeldovich, in Moscow, independently showed by a series of systematic arguments that the Universe contains more matter than antimatter. In particular, they noted, the primary cosmic rays striking Earth from every direction in space are predominantly made up of matter particles — mainly protons. Only about one in every 10,000 cosmic rays entering the upper atmosphere is an antiproton, and even that tiny number can be reasonably explained as due indirectly to the interactions of matter. An energetic proton can occasionally strike another proton in interstellar space. If the collision energy is high enough, a *third* proton plus an antiproton may be created, and all four particles will shoot away in various directions. The abundance of antiprotons expected at Earth from such encounters is about the one in 10,000 actually witnessed.

Except for this tiny contamination, therefore, cosmic rays "sample" the matter and antimatter content of the nearby Universe. Lower energy cosmic rays, which come predominantly from our own galaxy, confirm that there are no hidden pockets of antimatter within the Milky Way. High-energy cosmic rays, by contrast, come predominantly from *other* galaxies because the magnetic fields in our galaxy are simply not strong enough to trap such fast-moving particles. They probably originated in nearby galaxies like M31, or in the thousand or so galaxies of the Virgo cluster (of which M31 and the Milky Way are members). So we know that the whole Virgo cluster, at least, is made of matter.

There is another reason why there can be no antimatter in the Virgo cluster. We see no evidence anywhere in the entire system for the violent reactions that would occur if antimatter came into contact with matter—which must occasionally happen if the cluster contained any substantial quantity of antimatter. When a proton and an antiproton collide, they can produce high-energy photons, called "gamma rays," with specific energies. Although astrophysicists have searched for such photons in many directions, using gamma-ray telescopes borne by Earth satellites, they have not found any unusual emanations except for those originating from the center of the Milky Way. Such outbursts can be explained by conventional means, however, involving the high density of cosmic rays coming from the galactic center. In every other direction, there is no evidence for antimatter. On the basis of such observations, we conclude that the Virgo cluster contains only matter.

Another case against antimatter, which applies to all distances and directions, comes from our solid understanding of Big-Bang nucleosynthesis. If the Universe were symmetric in both matter and antimatter, then the vast numbers of particles and antiparticles would have completely annihilated one another—electrons with positrons, protons with antiprotons, and so on—leaving only scant traces of each, plus an absolutely tremendous number of photons. These annihilations would have ceased only when the remaining supply of particles and antiparticles had become so small that they hardly ever encountered one another any more. This would have been the case, however, when there was only one baryon left for every billion billion photons, but a ratio of about one in a billion is needed for nucleosynthesis of helium to occur in the proportions observed today. With such a huge number of photons flying around, complex nuclei could not have formed until much later in the era of nucleosynthesis, and there would have been far less helium produced. We are therefore confident that there was a small excess of matter over antimatter by the time the Universe was 1 second old and nucleosynthesis began.

If the Universe had been perfectly symmetric, in the last analysis, the few remaining unannihilated baryons could not have provided anywhere near the mass we witness today in the luminous regions of galaxies. Just about everything in existence would have ended up as a soft glow of microwave radiation. The very ground beneath our feet and the stars above our heads testify to the fact that we live in a Universe made of matter.

———————————— • ● • ————————————

So how did the excess matter arise in the first place, if the known processes of physics did not produce it? A possible answer to this puzzle, although not a fully detailed solution, was suggested in 1967 by the late Soviet physicist Andrei Sakharov, designer of the Soviet H-bomb and outspoken champion of human rights in the USSR. An excess of matter could occur, he proposed, if there were some kind of (as-yet undiscovered) interaction that did not conserve the number of baryons — that could lead, say, to a net *increase* in the number of protons in the Universe. (Physicists say that a reaction "conserves" something when the amount of it does not change during the reaction.) While this may seem like a tautology, it was actually a radical insight that was ten years ahead of its time. Physicists ignored Sakharov's proposal until the late 1970s because all forces then known appeared to conserve baryons.

Actually, Sakharov's proposal had additional provisos. Whatever force was involved had to violate a principle called "CP symmetry." Briefly put, CP symmetry means that the laws of physics remain the same if you replace all particles in an interaction by their antiparticles and view the process in a mirror. A violation of this symmetry principle is equivalent to a statement that matter and antimatter do *not* behave in exactly the same way.

In 1963, however, a violation of CP symmetry by the weak force was discovered by James Cronin and Val Fitch of Prince-

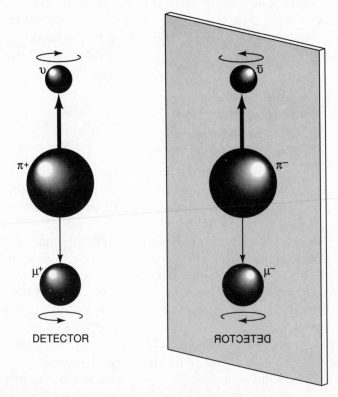

CP symmetry as observed in the decays of pions. If you replace all particles by their antiparticles and consider the "mirror-image" process, the rate of decay remains the same.

ton. Studying the decays of neutral kaons, they observed a small fraction (less than 1 percent) that did not obey the principle of CP symmetry. With experimental evidence that the weak force did indeed violate this symmetry, it took no great leap of imagination for Sakharov to propose another force that could do the same and also generate an excess of baryons in the process.

A final requirement was that this mysterious CP-violating force had to go out of equilibrium at some point in the first hundredth of a second. In a state of equilibrium — be it chemical, nuclear, or whatever — reverse reactions must proceed at exactly the same rate as forward reactions. Under such conditions, any asymmetry that might have built up between matter and antimatter due to some interaction would be immediately wiped out by the reverse process. To preserve a primordial asymmetry and lock it into existence forever, Sakharov had to insist that his hypothetical interaction occur *out* of equilibrium. The forward reaction responsible for a baryon excess had to occur faster than its reverse.

But a force that did not conserve the number of baryons was a fairly heretical idea in 1967, and Sakharov's proposal fell mainly on deaf ears. Weinberg made no mention of it in *The First Three Minutes*, even in his chapter about the first hundredth of a second. By the late 1970s, when such a force finally began to be taken seriously, Sakharov's idea had been all but forgotten.

———————————— • ● • ————————————

The late 1960s and early 1970s were a time of great upheaval in the field of elementary-particle physics. Not only were quarks discovered during this period, but unified theories of the forces between elementary particles also began to appear and be borne out by experiments. Such ideas held that the strong, weak, and electromagnetic forces were different manifestations of a single fundamental force. These striking theories gave cosmologists the tools they needed to pry back into that first hundredth of a second and catch a fleeting glimpse of what happened then.

Weinberg had a major role to play in the first step of this unification process. A visiting professor at MIT in 1967, he elaborated upon an idea advocated earlier by Harvard's Sheldon Glashow — his old friend from high-school days and an intellectual rival ever since. Glashow had suggested that the weak and electromagnetic forces might be just two different aspects of one and the same force. Weinberg came up with a specific process by which such an "electroweak" unification might occur. The Pakistani theorist Abdus Salam, working independently at London's Imperial College of Science and Technology, made a similar proposal the very next year.

The electroweak theory required that there be an extremely heavy, neutral particle — the Z particle — to act as an additional carrier of the weak force. Indirect proof of its existence came swiftly, at CERN in 1973 and Fermilab in 1974, setting the stage for the great surge of interest in unified theories that has been occurring ever since. Glashow, Salam, and Weinberg shared the 1979 Nobel prize for their ephochal prediction. Five years later Carlo Rubbia and the CERN accelerator physicist Simon van der Meer were awarded this prize for their roles in the 1983 discovery of the massive Z, the final and ultimate proof of electroweak unification.

A unification of two supposedly fundamental forces is not to be taken lightly. The last occurrence of such a feat had come a century earlier, when the Scottish physicist James Clerk Maxwell combined electric and magnetic forces into what he called the "electro-magnetic" force. Almost 200 years before that, Isaac Newton had shown that apples fell to the Earth and moons and planets stayed in orbit due to one and the same force — gravity. Albert Einstein spent the last 30 years of his life trying to unify gravity with electromagnetism, but to no avail.

Following the 1973 confirmation of the electroweak theory, Glashow and Salam (together with their colleagues Howard Georgi and Jogesh Pati) independently began proposing even more sweeping theories called "grand unified theories," or

"GUTs" for short. These theories hold that the strong, weak, and electromagnetic forces are just the different guises of a single fundamental Force—much as ice, water, and steam are different forms of H_2O. At the low energies encountered on Earth (or even in most particle accelerators), these three forces behave differently. But at the extremely high energies that should have occurred during the first split second of the Big Bang, they would have been essentially identical and have acted upon particles with much the same vigor.

GUTs are in fact an entire class of theories—of which Georgi and Glashow's (denoted SU5) is just the simplest version. A common feature of all GUTs is that the total number of baryons in the Universe can actually *change*. Remember, this was the key to Sakharov's 1967 proposal. At the stupendous energies where the three forces become unified, according to these theories, quarks and leptons lose their differences too. A quark can readily change into a lepton, or vice versa.

Actually, the same processes can occur even at the low energies and temperatures prevalent today, but much more slowly. If one of the quarks inside a proton converted into an electron, for example, the proton would disintegrate. All grand unified theories predict that a proton or neutron can spontaneously decay into a state involving no baryons at all, although this process should take at least 10^{32} years, on the average. That's about the time it would take an old man to wipe down a very tall mountain by stroking it once every 100 years with a soft cloth!

For most of the 1980s, experimental physicists sought to detect such proton or neutron decays by patiently observing thousands of tons of water or iron sitting deep underground, where they were shielded from cosmic rays. With more than 10^{33} protons and neutrons in these detectors, there was a good chance of witnessing such decays—if the GUTs were correct. But none have been observed yet, which casts serious doubt on Georgi and Glashow's original SU5 theory. Because of this lack of experimental proof, the status of grand unified theories must be viewed as conditional, no matter how appealing they might be to theorists.

Nevertheless, a majority of cosmologists now take it for granted that *some* form of grand unified theory must have been in force at the extremely high energies encountered in the very early Universe, well before it was one-hundredth of a second old. The trend of the strong, weak, and electromagnetic forces at currently accessible energies seems indeed to point toward such a unification. And GUTs provide a natural way to explain why there is so much matter in the Universe — and not just an all-pervasive bath of microwave radiation.

———————————————— • ● • ————————————————

So let's use some of these recent insights from particle physics, both established dogma and informed speculation, and see what they can reveal about the first hundredth of a second. Try to imagine you are watching a movie of that brief period, but it is rolling back in time frame by frame. As the film proceeds backwards, the particles get closer and closer together, and the temperature climbs steadily higher as their average energy grows.

At 0.01 second, the temperature is 10^{11} degrees, corresponding to an average energy of about 10 MeV per particle. The Universe is a hot broth of matter and radiation, populated about equally with photons, electrons, positrons, neutrinos, and anti-neutrinos. As mentioned above, its density is a billion times that of water. There is a sprinkling of protons and neutrons in this soup, about one for every billion photons — just the leftover residue of earlier activity. Antiprotons and antineutrons are now virtually nonexistent. Although the temperature is high, it is not high enough to keep recreating these heavy particles every time they are annihilated.

These conditions persist as the film rolls backwards. Everything gets steadily hotter and denser, reaching a trillion degrees at 0.0001 (10^{-4}) second. At slightly earlier times, when the average energy per particle exceeds 100 MeV, we begin to see new kinds of heavy particles created in tandem with their anti-particles. The first to appear are the muons, heavy leptons with a mass of 105 MeV, about 200 times as heavy as the electron. With still higher temperatures come pions and kaons, the mesons that serve to carry the strong force between baryons. Here is the "melange" that Weinberg lamented, a complex soup that still gives theorists fits today.

When the temperature nears 10^{13} degrees at an age of about 10^{-6} second, or 1 microsecond, something remarkable occurs. The temperature is now so hot, several hundred million electron volts per particle, that the mesons and baryons begin disinte-

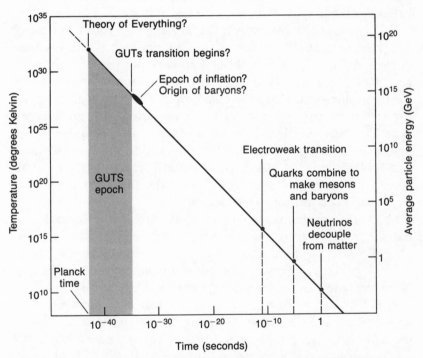

The particle history of the Universe. As its temperature falls steadily, the Universe passes through several transitions from one epoch to the next; different kinds of particles and interactions dominate in each epoch.

grating into their elementary constituents — quarks and anti-quarks. At the moment the quarks dissociate, we begin to encounter free gluons, too. These, you may recall, are the elusive carriers of the strong force between quarks.

At times earlier than 1 microsecond, the Universe was a hot broth of free quarks and leptons. Photons and gluons flitted incessantly among these particles, bearing the electromagnetic and strong forces between them. The behavior of this soup is well understood. The Standard-Model theories of quantum electrodynamics (QED) and quantum chromodynamics (QCD), which describe the behavior of photons and gluons, enable us to make calculable, verifiable predictions when quarks and leptons are so close to one another. (Only when the quarks begin to get farther apart, as occurred after about 1 microsecond, do the calculations of QCD become extremely difficult.)

Even when the temperature reaches 10^{15} degrees at a time of 10^{-10} second, the behavior of this quark–lepton soup is well known. At energies above about 100 GeV per particle, the massive W and Z particles come into their own and appear abundantly in the soup. Thanks to Glashow, Salam, and Weinberg, we can calculate the behavior of these particles with great precision. These calculations have been repeatedly checked and verified, using the highest energy particle colliders now available at CERN, Fermilab, and SLAC, which actually reproduce on a small scale conditions that occurred during the Big Bang at one-trillionth of a second, or 10^{-12} second.

So far our movie has been based on hard, experimental facts gained by simulating the conditions of the Big Bang at earthbound particle colliders. As we roll our film back beyond 10^{-12} second, however, the script begins to become more speculative. Now we enter a domain where particle energies exceed those producible on Earth, and we have to trust that our theories are actually valid at these ultrahigh energies. Is what actually happens under these extreme conditions a smooth extrapolation of the physics known to occur at lower energies?

There are a number of indications and arguments that the

answer is indeed yes, but we cannot be absolutely sure. Nature has often had a surprise up her sleeve for unwitting physicists who made such blithe assumptions in the past.

But let's make that leap anyway and try to discern what might have occurred in that first trillionth of a second, based on informed speculation about physics at ultrahigh energies. Some of these speculations may soon be tested experimentally as ever more energetic particle colliders, such as the Superconducting Super Collider being built in Texas, begin operating. If the simplest grand unified theories are correct, there are no surprises awaiting us until we work our way back to 10^{-34} second, when the average energy per particle has reached a truly astounding 10^{14} GeV. That's almost a million billion times the mass–energy of a proton. (Actually, many theories predict that new particles would appear at lower energies, too, but these entities would not alter this basic picture.)

According to grand unified theories, and remember that there are many different versions, we should then begin to witness the appearance of stupendously massive particles called X particles, with masses of 10^{14} to 10^{16} GeV each. To put these masses in perspective, this is about as heavy as a good-sized speck of dust containing many trillions upon trillions of atoms. If a single high-energy X particle smacked you in the face, you'd certainly feel its impact!

In the GUTs epoch, as the period before 10^{-34} second is called, the strong, weak, and electromagnetic forces acted as a single unified force, and quarks could become leptons (or vice versa) through the intermediation of these X particles. A quark disgorges an X, for example, as it converts into an antilepton; the X subsequently decays into a quark and a lepton. Because X particles are so terribly massive, such processes would not occur (or would occur extremely rarely) at normal temperatures. At the ultrahigh temperatures and energies encountered during the GUTs epoch, however, X particles would be almost as abundant as photons, and these processes would occur all the time.

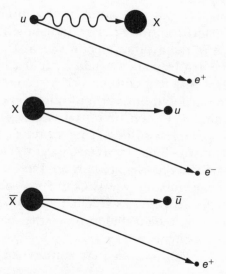

Interactions of X particles with quarks and leptons. According to the grand unified theories, these stupendously massive carriers of the GUTs force would have been commonplace before 10^{-34} second.

Having rolled our film of the Big Bang back to this very early instant, let us pause and then roll it forward again. The stage is set for the creation of matter.

• ● •

In a Universe populated with X particles, it is possible to satisfy Sakharov's three requirements and create more matter than anti-matter. Remember that he needed a force that could alter the number of baryons and violate CP symmetry—and do so out of equilibrium. The first person to recognize this possibility, in 1978, was Motohiko Yoshimura, a Japanese particle theorist who had become interested in cosmology while studying at the University of Chicago. But he unfortunately missed the third requirement, that the Universe somehow had to go out of equilibrium, a point that was soon made by Weinberg and others. Only later was Sakharov's prior work rediscovered.

To see how such an asymmetry can arise, imagine we start with equal numbers of X and anti-X particles, as we would expect to encounter in a symmetric Universe. The X decays yield quarks and the anti-X decays yield antiquarks. Now if CP symmetry is violated by the GUTs force that governs these decays, it becomes possible to generate a small *excess* of quarks over antiquarks—more matter than antimatter! Remember CP symmetry means that matter and antimatter behave in essentially identical ways. So if this symmetry is violated, as happens in grand unified theories, we can obtain more of one than the other.

If the Universe happens to be in equilibrium, however, the reverse reactions proceed with equal strength and immediately wipe out any slight excess that crops up. Quarks give back X particles; antiquarks recreate anti-X particles. This is the crucial point that Yoshimura missed and others recognized. It is only when the Universe goes out of equilibrium, at the close of the GUTs epoch, that a *permanent* asymmetry between the quarks and antiquarks can ever be established. (Remember, we are running our film forward now.) At this point the temperature has fallen so far that there is not enough energy available to recreate X and anti-X particles as rapidly as before; they begin to decay faster than they can reappear. Now we can indeed achieve a tiny excess of quarks, a permanent residue of matter frozen into the Universe forever.

The excess amount of matter remaining after this epoch depends on the actual details of the particular grand unified theory we invoke. But it is important here to remember that *all* GUTs meet the first two of Sakharov's requirements—nonconservation of baryons and violation of CP symmetry. The cooling of the Big Bang always gives us the third requirement, a nonequilibrium condition at the end of the GUTs epoch. The creation of matter is inevitable in GUTs; the only question remaining is how much matter is produced.

As our film rolls forward after the GUTs epoch, and the Universe continues expanding and cooling, quarks and antiquarks eventually begin to coalesce into baryons and antibaryons. This condensation process begins at a time of about 1 microsecond, when the average energy per particle has fallen to a few hundred million electron volts. Because we started with a slight excess of quarks, we finish with a tiny excess of baryons. In the ensuing orgy, baryons and antibaryons annihilate one another to yield vast quantities of photons; a tiny fraction of baryons remain — those that simply could not find an antibaryon with which to annihilate. This remnant becomes the matter we witness in the Universe today — if the GUTs scenario is indeed correct. Grand unified theories provide a natural way to generate an excess of matter over antimatter, even if the Universe began in a state of perfect symmetry.

———————————— • ● • ————————————

An important relic of that first hundredth of a second is the number of baryons remaining. This quantity is usually expressed as a ratio of the number of baryons to photons in the present Universe, which we will denote as B/P. This key ratio must lie between 0.2 and 0.7 parts per billion if Big-Bang nucleosynthesis is to yield the observed abundances of light elements. That's a pretty tight constraint. The various GUTs yield different values of B/P, as has been demonstrated by groups of physicists at CalTech, CERN, Chicago, and Stanford. So the observed values of this ratio can be used to rule out some of the possible theories. Georgi and Glashow's original SU5 theory, for example, fails to account for B/P unless additional assumptions are included. Here, once again, cosmology has revealed information about fundamental physics.

In fact, the SU5 theory has encountered other difficulties, and has fallen out of favor among particle physicists and cosmologists. The proton decay rate expected from SU5 was not observed in underground detectors, as noted earlier, and the theory predicts too small a value for the Weinberg angle, a key parameter in the unified electroweak theory that indicates the relative strengths of the electromagnetic and weak forces. Robert Marshak, a theorist from Virginia, was sufficiently dissuaded by the evidence to pronounce SU5 officially "dead" in 1986 at a major international conference on high-energy physics.

The idea of grand unification itself, however, is still healthy. There are many other possible GUTs, possessing other kinds of symmetry, that do not have the problems of SU5. Perhaps the most interesting of these are the theories that incorporate what is known as "supersymmetry," or "SUSY" for short. We will say more about supersymmetry later in this book; here let it suffice to note that all supersymmetric theories effectively *double* the number of fundamental particles by introducing a partner to each known as a "sparticle," or "SUSY particle." The partner of a quark is called a "squark," that of a lepton is a "slepton." Photons have "photinos" for partners, gluons have "gluinos," W's have "winos," and so on. It is not important to remember all these names, only to realize that an entire new realm of elementary particles may be just around the corner. Some of the SUSY particles, in fact, are prime candidates for the dark matter in the Universe.

In supersymmetric versions of grand unified theories, a number of irritating problems and anomalies are cleared up. For one, the proton lives much longer and disintegrates into combinations of particles that would probably not have occurred thus far in underground detectors. This feature would explain why proton decay has not yet been observed. The electroweak parameter known as the Weinberg angle comes out correctly in certain SUSY GUTs, too, and the baryon-to-photon ratio B/P can be brought into accord with what is needed for Big-Bang nucleosynthesis to occur.

There is growing realization today among theorists that *some* form of supersymmetry must be included in particle theories in order to obtain a complete and correct description of the early Universe. So the additional complexity of these supersymmetric

theories, with twice the fundamental particles previously required, may be a necessary evil dictated by the hard realities of
observation and measurement. Particle physicists are eagerly
searching for these SUSY particles at the new colliders that have
recently begun operations.

———————————— • ● • ————————————

An intriguing aspect of grand unified theories, or of *any* theory
that generates matter in the early Universe, is that it may govern
how "bumps" could have arisen in the original distribution of
matter. These primordial density fluctuations, however small,
would have acted as "cosmic seeds" upon which all matter soon
began clumping, eventually leading through several stages of
evolution to the formation of galaxies, clusters, and superclusters—and stars, planets, and people.

Had the distribution of matter been absolutely smooth everywhere, this process of clumping could never have begun, and
galaxies and clusters would not exist. The fact that we see them
almost everywhere we gaze in the skies above suggests that there
must have been some sort of density fluctuations or inhomogeneities when matter was first created—or some other kind of
cosmic seeds that induced these structures to begin forming.
They are yet another relic of conditions in the early Universe.

There are two distinct ways in which bumps could appear in
the early distribution of matter, either by "isothermal" or by
"adiabatic" fluctuations. In the first, matter is clumped a little
but radiation (the sea of photons, that is) is distributed perfectly
smoothly. Such a fluctuation is said to be isothermal because the
radiation must have a uniform temperature everywhere if it is
absolutely smooth. In isothermal fluctuations, therefore, the
ratio of matter to radiation can change from place to place if
there is any lumpiness in the distribution of matter. In adiabatic
fluctuations, by contrast, both matter and radiation clump together at the same places simultaneously. Whenever matter gets

A rich cluster of galaxies, known as the Coma cluster, in the constellation Coma Bernices.

denser, that is, so does the radiation. But the *ratio* of matter to radiation remains the same everywhere, even though either quantity can differ from one point in space to the next.

Prior to the late 1970s, when theorists began to realize how matter could have been created within the context of grand unified theories, isothermal and adiabatic fluctuations were considered equally plausible. But any microscopic mechanism for making matter in the early Universe (such as GUTs) requires that radiation and matter be intimately linked with each other. Matter could have been created when the GUTs force fell out of equilibrium, which can only occur at a specific temperature. If the temperature of the early Universe were exactly the same everywhere, the distribution of matter would consequently have ended up perfectly smooth. If there had been slight differences in temperature from one point to the next, however, there would also have been bumps in the matter density. Grand unified theories therefore favor *adiabatic* fluctuations: any variations in matter density have to be accompanied by corresponding variations in temperature.

This correlation between density and temperature is extremely important, because precise measurements of the microwave background radiation can detect very slight variations in its temperature. Accuracies of a part in 100,000 are achievable today (as we shall discuss in the next chapter), and improvements by another factor of 10 may be possible soon. As yet, however, no significant variations in the temperature of this radiation have been found. So any initial fluctuations in matter density must also have been extremely small, if we assume they had to be adiabatic.

———————————————— • ● • ————————————————

Although we can now explain the creation of matter, a different puzzle begins to loom on the horizon. On the one hand, we observe galaxies and clusters everywhere we look, so there must have been some kind of cosmic seeds upon which matter began collecting — be they primordial density fluctuations or whatever else. On the other hand, the microwave radiation coming from all directions in space appears to be extremely smooth, as far as we can tell today. The isothermal fluctuations that could explain such a difference are not easily compatible, however, with grand unified theories, the current favorite explanation for the creation of matter.

What is this apparent contradiction telling us? Is there something missing in our picture of the early Universe?

6

The Birth of Galaxies

*L*ife as we know it depends upon a number of physical "accidents" that might seem miraculous to a casual observer. Were the force of gravity slightly stronger than it is, for example, the Sun would have expired billions of years ago — long before vertebrates ever had a chance to appear on the surface of its watery third planet. A stronger gravity would have squeezed the Sun more tightly, elevating its core temperature; this increase would have driven its thermonuclear fires much faster and burned up its hydrogen supply far earlier. Were gravity much *weaker*, on the other hand, our nearby star would be too cool to support any life on Earth at all. Its core temperature would be too low for any significant nuclear activity to occur.

Another apparent miracle is the extreme density of matter in our immediate surroundings. The many complex chemical reactions that created and sustain us, fueling our bodies and minds, are impossible in the seeming emptiness of space that is the norm throughout the Universe. Such reactions can occur only in regions such as Earth where there exist tremendous numbers and great varieties of atoms in extreme proximity.

If all the visible matter in the Universe were distributed uniformly throughout space, however, there would be only a few baryons in every 100 cubic meters. That's the equivalent of a single hydrogen atom rattling around in a big trailer truck. Even if we included all the invisible matter, there could be no more than one baryon per cubic meter, on the average. To appreciate how sparse even this much matter is, imagine that a hydrogen atom has been blown up to the size of a mustard seed. Its nearest neighbor would be 10,000 kilometers away, or almost halfway around the globe!

Nearby space in the solar system contains much more matter than this. Here on Earth, a cubic meter of ordinary material — water, sand, or wood — contains in the neighborhood of 10^{30} (or 1,000,000,000,000,000,000,000,000,000,000) baryons; we say it is about 30 "orders of magnitude" denser than average. Because there are some exceedingly dense regions in the Universe, then, there must also be vast regions of incredible emptiness in between, to make the overall density average to one baryon per cubic meter. How did such a state of lumpiness ever arise?

From the smooth distribution emerging out of the Big Bang, gravity must eventually have clumped matter together into tiny pockets of extreme density, leaving behind enormous voids in space. Individual galaxies must have coalesced and later fragmented into stars and planetary systems. To interpret this visible structure, we need to understand how these clumps began forming from the initially uniform conditions — and how they have evolved to their present configuration.

Getting matter to begin clumping at all is a difficult task, given how smooth the Universe must have been at its beginning. From the uniformity of the cosmic background radiation, we know that matter must have been very evenly distributed when the Universe was about 100,000 years old — the moment radiation and matter parted company. Thus, normal matter, which was in equilibrium with radiation in a single hot plasma until that time, must have been smoothly distributed, too. Before that instant, any small ripples arising in this sea would have been quickly smoothed out to conform with the radiation. Only *after* their divorce could such ripples grow into larger clumps, eventually forming the galaxies and clusters that are the dominant visible features of the structure of the Universe today.

———————————— • ● • ————————————

A principal relic of the moment when matter escaped from light, thus allowing structures to begin forming in the early Universe, is the cosmic background radiation. From its apparent temperature and its distribution across the sky, we can reach a number of important conclusions about this period. Indeed, this soft afterglow of creation is one of our major pieces of evidence for the hot Big-Bang model of the Universe. How did it arise?

A sudden divorce occurred when the Universe was around 100,000 years old, changing things irrevocably. Just before that time it was a hot plasma of unbound electrons and protons—plus helium and a sprinkling of other nuclei that had been formed during the era of nucleosynthesis. These particles of matter were awash in a vast sea of photons, or electromagnetic radiation. Matter and radiation were in intimate contact, much like a school of sardines swimming in the ocean.

When the temperature fell to about 3000°K, however, the negatively charged electrons no longer had enough energy to remain free; they began sticking to the positively charged nuclei, forming electrically neutral hydrogen and helium atoms. Thus confined, these charged particles were no longer able to interact with radiation—at least not very readily. Photons finally burst free of matter and could travel unimpeded through the thin gas of the early Universe, which had suddenly become

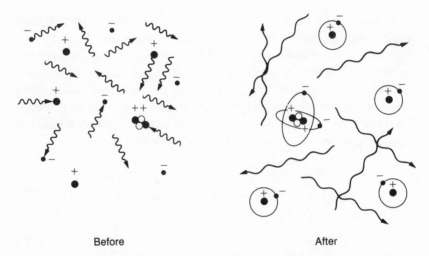

Before   After

The decoupling of matter and radiation, which occurred when the temperature of the Universe had fallen to about 3000°K at an age of about 100,000
years. After that event, photons (depicted as wavy arrows) could travel unimpeded for great distances.

as transparent as air. In effect, matter and radiation parted company. Scientists say they "decoupled."

A well-known property of matter is the propensity of any
continuous surface — such as a wall or floor — to emit electromagnetic radiation. The wavelength of this radiation depends
only on the surface temperature of the emitting body, not its
composition. At a temperature of about 5000°K the Sun radiates
photons mainly at optical wavelengths. (In fact, our eyes probably evolved to detect such photons because they were the dominant radiation available.) People radiate at longer, "infrared"
wavelengths, which our eyes cannot detect, but sniper scopes
and other infrared sensors can. The pages of this book give off
slightly longer infrared wavelengths because books are generally
cooler than people. The cooler an object, the longer the wavelength of its emissions.

When the Universe was still a hot plasma at the 100,000-year
mark, its temperature was roughly the same as the surface of the
Sun today. As a single continuous body at this one temperature,

it radiated profusely in the same range of optical wavelengths. Just before escaping from matter, this radiation was predominantly composed of visible and shorter wavelength ultraviolet light. After breaking free, however, the radiation could not leave the Universe, which contains everything that exists. The remnants of this radiation therefore must still be around today.

But because space has expanded tremendously since that early time, the wavelengths of the leftover radiation must have increased to the point where they are far longer now than when it was first emitted. Think of this radiation as stiff springs whose coils were once tightly wound but have since been stretched so much that there are large gaps between successive turns. Although short-wavelength visible light was emitted by matter at about 3000°K, it reaches us now at far longer microwave and radio wavelengths — as if it had originated from a body with a uniform temperature of about 3°K. That's the kind of dull glow we expect to observe from an object cooled to the very cold temperature of liquid helium.

It was the detection of this radiation in 1964 that set the stage for modern Big-Bang cosmology and won Arno Penzias and Robert Wilson the 1978 Nobel prize in physics. As noted in Chapter 1, these two scientists at Bell Laboratories used an existing radio antenna in Holmdel, New Jersey, to discover a uniform glow of microwave radiation striking Earth from every direction. Their serendipitous discovery was quickly confirmed by another group of scientists at nearby Princeton University, who had been building a radio antenna designed to search for just such an effect.

Others who had used the Holmdel antenna previously also witnessed the same soft glow, but had unthinkingly set the zero of their measurement scale to the top of this "background" and searched only for bumps above it. Having read the earlier accounts (showing no uniform microwave radiation), the Soviet theorist Yacov Zeldovich had even concluded that the hot Big-Bang idea of Gamow and colleagues was in fact wrong. Unlike their predecessors, Penzias and Wilson recognized that the "noise" level in their equipment was far too high, and spent months of hard work trying to eliminate or at least understand this curious effect. Their meticulous efforts were eventually rewarded by one of the most important discoveries in modern physics.

-------------------------------- • ● • --------------------------------

The striking uniformity of this relic microwave radiation indicates that the Universe was extremely smooth when the light was emitted about 15 billion years ago — when it was about 100,000 years old. No matter in which direction we look, the effective temperature of this radiation is the same. The only small difference from absolute uniformity established by 1991 is due to the Earth's motion relative to the rest of creation. After we remove that variation, the early Universe appears to have no ripples in its density greater than a few parts in 100,000. The surface of a billiard ball, for example, is much rougher than this. By some means, therefore, galaxies and clusters were able to evolve from what were extremely smooth initial conditions to the state of extreme clumpiness we witness today.

The detailed measurements of this relic radiation are a fascinating episode in radio astronomy. Whereas a little radiation can be detected at centimeter wavelengths using ground-based equipment like the Holmdel antenna, most of it reaches the Earth in the millimeter range, and these shorter wavelengths cannot penetrate the atmosphere. To look for minuscule variations in any signal, be it starlight or radio waves, astronomers need lots of signal. They have built radio antennas on mountaintops and lofted instruments in balloons or on U2 aircraft to altitudes near 30 kilometers — where the absorption of this microwave radiation by air is much diminished, and better measurements are possible. These studies have occurred all around the globe. One balloon came down in the Brazilian jungle and was presumed lost, only to turn up months later in a village bar, where it had been hung as an exotic decoration.

The U2 aircraft used by Berkeley and NASA physicists to sample the cosmic background radiation.

The first important task, after Penzias and Wilson's discovery had been confirmed by the Princeton group, was to prove that this uniform cosmic radiation had a spectrum appropriate for a continuous surface at about 3°K. For that to hold true, the radiation intensity had to increase as one moved from centimeter to millimeter wavelengths, peaking at about 1 millimeter and then dropping off sharply at shorter wavelengths. But making measurements at wavelengths around 1 millimeter is impossible to do from mountaintops; such a feat requires satellites, rockets, or extremely high-flying balloons.

The initial, rocket-borne measurements came in showing much higher intensities than expected, giving hope to steady-state cosmologists. Fred Hoyle had developed a new version of his theory that could reproduce the spectrum already observed

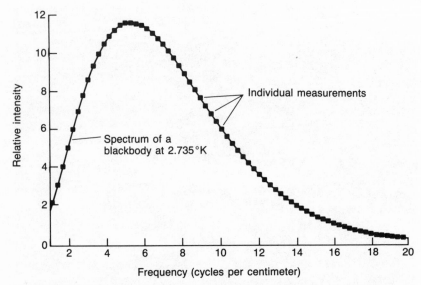

The spectrum of cosmic background radiation measured by the Cosmic Background Explorer, or COBE, satellite in late 1989. The measured data points (squares) can be fit almost perfectly by the spectrum emitted by a blackbody at a temperature of 2.735°K.

at centimeter wavelengths, but it would not give the sharply peaked spectrum predicted by the Big-Bang theory in the millimeter range. His hopes died soon thereafter, when astronomers realized how hard it was to measure low temperatures accurately with an instrument riding on a hot rocket.

By the mid-1970s, however, a group from the University of California at Berkeley led by Paul Richards had performed a series of balloon-borne experiments which finally proved that the radiation intensity indeed peaked near 1 millimeter, as expected from a body at about 3°K. Recent very accurate measurements by the Cosmic Background Explorer (or COBE) satellite have determined this temperature to be 2.735°K. The exact temperature is determined from the position and shape of the microwave spectrum; a lower temperature means the spectrum peaks at longer wavelengths or lower frequencies.

———————————— • ● • ————————————

In present-day searches for variations in the microwave background radiation, astronomers look for small discrepancies in the temperature of the sky between observations in two different directions. They seek a slight "dappling" in its otherwise uniform "color." An experiment may involve using a single device and repeatedly switching its orientation back and forth from one direction to another, trying to discern small shifts in the microwave spectrum. Or one might use identical antennas pointed in two or more directions at the same time. One astronomer may elect to study radiation from the vicinity of the constellation Hercules, for example, and then rotate his instrument by 180 degrees to look the opposite way. Another might aim her two antennas at two points in Orion only a few degrees apart. The physical interpretation of any temperature difference depends on the angle between the two measurements.

The only discrepancy established by 1991 is the so-called "dipole" variation. Measurements made from U2 aircraft and on balloon flights by groups from Princeton, Berkeley, and the University of Rome have shown that there is a well-defined difference of 1 part in 1000 between two specific directions that are 180 degrees apart. In one direction the microwave radiation is slightly redshifted; in the opposite direction it is shifted about an equal amount toward the blue end of the spectrum (to shorter wavelengths).

This small variation results from the motion of the Earth itself through the cosmic background radiation. When corrected for Earth's motion relative to the Sun, and the Sun around the Milky Way, it means that our galaxy must be moving at a speed of about 600 kilometers per second relative to the rest of the

Universe. That's 1.5 million miles per hour! We are rushing headlong toward two nearby superclusters — Virgo and Hydra-Centaurus.

To determine this motion accurately required mapping temperature variations over the entire sky, including both the northern and southern hemispheres. You can imagine the difficulty scientists had in convincing certain governments to permit U2 overflights for these measurements. Wary officials had to be assured that instruments would point only up, not down!

Variations in the sky temperature at angles less than 90 degrees apart can tell us about the smoothness of the early Universe, at the moment radiation and matter parted company. Small ripples in the density of matter at that time might have served as cosmic seeds that eventually grew into the visible galaxies and clusters we witness today. By 1991, no variations had been convincingly established at such angles, but astronomers had set stringent limits. David Wilkinson and his Princeton colleagues, for example, have published a limit of 2 parts in 100,000 for the possible temperature variation at angles of 7 arc minutes (60 minutes is 1 degree of arc). For measurements 90 degrees apart, the Soviet RELIC satellite has established a limit of at most 3 parts in 100,000 on the so-called "quadrupole" variation.

The limits are not quite as restrictive at angles between these two extremes, but no significant variations have been found at the level of 1 part in 10,000. Here is where the COBE satellite is making a major contribution. In the airless void of space, a satellite is the ideal platform from which to study this microwave radiation. Although it was designed in the mid-1970s, this satellite was only launched in late 1989 — due mainly to shifting priorities at NASA and the agency's emphasis on manned spaceflight versus unmanned probes. By January 1990, COBE had produced its first important data — the precise spectrum of cosmic background radiation shown on page 122.

An important goal of the COBE mission is to search for very subtle variations between separate measurements of this radiation made in two different directions. The angle between them

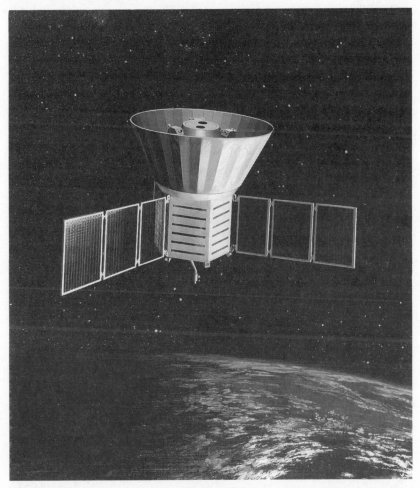

The COBE satellite, launched into Earth orbit in 1989, has measured the cosmic background radiation with unprecedented accuracy.

can be anywhere from about 10 to 180 degrees. Already a COBE team led by Berkeley physicist George Smoot has made the best determination of the dipole variation at 180 degrees and established an upper limit of about 3 parts in 100,000 on the possible variations that might occur at angles between 10 and 90 degrees (see graph below). If there are any variations in this angular range that are as much as 10 times smaller, the COBE group should eventually be able to detect them, but such a feat will require careful analysis of a large body of data. Hard experience has shown that there are many subtle but spurious effects that can masquerade as an apparent temperature difference.

In the immediate future, new ground-based experiments proposed for the South Pole may be able to do even better than COBE. The extreme cold and very dry air prevalent there make ideal conditions for observing cosmic background radiation. These measurements may eventually be able to detect temperature differences as small as 1 part per million.

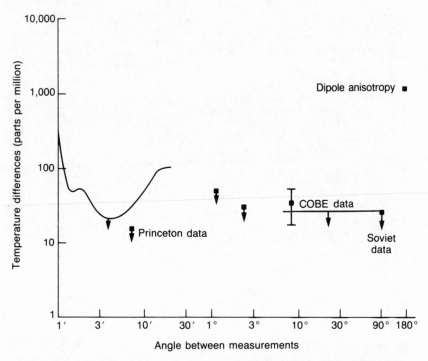

Limits on the difference between two measurements of the apparent temperature of the cosmic background radiation, plotted versus the angle between the measurements. So far the only confirmed difference is between measurements separated by 180 degrees—the dipole variation.

In making such measurements, scientists have to be extremely careful about spurious sources of microwave radiation. The Milky Way gives off radiation at many of the wavelengths examined in these experiments, so its contribution must be subtracted from the observed signals before any intrinsic difference can be established. One group had reported an apparent quadrupole variation (between measurements made 90 degrees apart) before they recognized it was just our galaxy radiating at unfamiliar wavelengths. The fact that various galaxies and clusters cover different portions of the sky make some angular scales more difficult to study than others, due to the extraneous radiation they contribute. When scientists try to measure such tiny differences as 1 part in 100,000 or smaller, they have to worry about all kinds of spurious effects. The COBE satellite has special equipment to look for such effects.

Because structures are observed on many different distance scales — galaxies, clusters, and superclusters — we expect that any primordial density ripples that led to these structures also occurred at many different scales, too. Edward Harrison at the University of Massachusetts and Zeldovich argued that these ripples were probably the same size on *all* scales, a distribution generally known as the "Harrison – Zeldovich spectrum." If true, we would expect to observe similar variations in the microwave radiation (once the dipole variation and all other extraneous effects are subtracted) at all angular separations between a few minutes and 90 degrees. Thus, according to Harrison and Zeldovich, the best limits established anywhere in this range should apply to the entire range as a whole. So they would have expected no variations in the microwave radiation greater than 2 parts in 100,000 — no matter what the angle between two measurements.

Whatever the case, any intrinsic variations are extremely small — less than 1 part in 10,000 at all angles, and a few parts in 100,000 in some cases. This relic of creation is an extremely smooth, uniform glow suffusing all of space.

In all likelihood, the smooth background radiation witnessed today means that matter was spread out very uniformly at the time it decoupled from radiation 15 billion years ago. Any ripples in this vast ocean must have been extremely small then — assuming, of course, that they were *adiabatic* fluctuations, in which matter and radiation traced one another. As we noted in the last chapter, these are the kind of fluctuations that occur naturally in grand unified theories, which offer our only decent explanation of why the Universe contains so much matter. Isothermal fluctuations, wherein the radiation is smooth while matter can be bumpy, can account for the uniform background radiation too, but seem incompatible with GUTs. (However, GUTs can also generate "defects" in the fabric of space itself, which can serve as cosmic seeds upon which structures grow; we will examine this option in later chapters.)

Suppose we began with a small ripple in the distribution of matter. How did it ever grow up to be a great big galaxy? How did such a fledgling fluctuation evolve over the intervening billions of years to become one of the magnificent structures we witness today? This is a very good question, particularly since our observations of the cosmic background radiation tell us that the original ripple must have been small indeed — like a foot-high wave on a 2-mile lake.

There are two distinct periods of evolution: an initial period of slow, "linear" growth followed by a period of rapid, "non-linear" compression. In the first period, gravity holds our aggregate loosely together while the rest of the surrounding matter thins out gradually with the general expansion of the Universe. The slight overdensity in this pocket of matter keeps it from expanding quite as fast as the rest of existence. There is enough matter in the nearby regions of space, however (all of it trying to yank our poor aggregate apart), that the density in this pocket can only grow incrementally.

Such a tense state of affairs continues until the density in surrounding regions has fallen to about half that inside the little pocket. At this point, the self-gravity of our clump takes over in earnest and triggers its swift collapse. The disruptive gravita-

Formation of a spiral galaxy from an overdense cloud of gas. A period of slow, linear growth is followed by rapid collapse of the cloud into a rotating disk of luminous matter.

tional pull of the surrounding material has by then weakened to the point where it can no longer keep the clump from contracting due to its own self-attraction. Once this process of contraction begins, it accelerates feverishly because higher densities lead to stronger gravitational forces, and vice versa. An extremely rapid growth in density ensues, eventually yielding familiar celestial objects like galaxies, stars, and planets. These can easily be 30 or more orders of magnitude denser than the universal average.

During the slow, linear growth period, the excess density of the clump (relative to the average) grows in inverse proportion to the temperature of the Universe. If the average temperature was 3000°K at the moment of decoupling and became 30°K at a later time, say 1 billion years afterward, then the overdensity would grow by a factor of 3000/30, or 100. If this excess was 1 part in 1000 at decoupling, it would become 1 part in 10 by the time the temperature reached 30°K. This is what is meant, in fact, by "linear growth."

Now remember that the linear growth period always continues until the excess density in the clump equals the average density in the Universe, a density enhancement of 1 part in 1. Only then can the rapid, nonlinear growth begin that yields the fantastic densities witnessed in today's Universe. Because the average temperature is about 3°K today and was 3000°K at the moment of decoupling, this means a density enhancement can have grown by only a factor of 1000, at the very utmost, during this linear growth period. So only those regions with an over-density of *at least* 1 part in 1000 have any chance of collapsing by the present day.

We may be witnessing the onset of such an event in a relatively nearby region of space. In 1989, Cornell University astronomers Riccardo Giovanelli and Martha Haynes discovered a slowly rotating cloud of hydrogen gas that seems on the verge of collapsing. About 70 million light-years away, this elliptical cloud is about 10 times as large as the Milky Way and contains about one-tenth as much matter. If it is indeed entering its phase of nonlinear growth, it must have sprung from an overdensity that was only about 1 part in 1000 at the moment of decoupling. Its linear growth phase has taken a whopping 15 billion years or so!

The high densities we witness today, about 30 orders of magnitude greater than the universal average, tell us that some parts of the Universe (including the one beneath our feet) must have entered their period of rapid collapse long before the present time. Thus, there must have been density enhancements much *greater* than 1 part in 1000 at the moment of decoupling, if this scenario is correct. An overdensity of 1 part in 100, for example, would have grown to 1 in 1 by the time the Universe had cooled to 30°K at an age of 500 million years. At that moment this clump would have initiated its collapse, eventually becoming a spiral nebula or a quasar (short for "quasi-stellar radio source"), in a relatively short time. In the past few years quasars have been identified with very large redshifts, indicating that their period of nonlinear growth must have occurred during the first billion years, at least for some of them. Many of the galaxies and clusters we see out in space, as we look billions of years back in time, must have been formed in similar ways.

But now we have a big problem. From the observations of the cosmic background radiation today, we know that matter was *extremely* smooth at the moment of decoupling. Any overdensities at that time must have been at most 1 part in 10,000 — and maybe as small as a few parts in 100,000. These ripples are far smaller than those necessary to trigger gravitational collapse before the present day. But the compact structures we witness in all directions tell us that such collapses occurred almost everywhere. What is wrong here?

----------------------------- • ● • -----------------------------

This seeming paradox has led cosmologists to propose several possible solutions, which follow two principal paths. In the first approach, they question the validity of the usual assumption that matter was distributed exactly like radiation during the early Universe. In *isothermal*, as opposed to adiabatic, fluctuations, you recall, the radiation can be smooth while matter is clumpy. In this way one can allow radiation to have bumps in it smaller than 1 part in 10,000, but matter to have the better than 1 in 1000 overdensities necessary to induce gravitational collapse by the present day.

But this approach flies in the face of the favorite scenario for matter creation using grand unified theories. In GUTs, as we mentioned in the last chapter, an excess of matter is created over antimatter at about 10^{-34} second into the Big Bang. Matter creation was a microscopic process that occurred just as the Universe fell out of equilibrium at the end of the GUTs epoch. It is difficult to understand how this new matter could be clumpy while the radiation that fostered it somehow remained smooth. (There is, however, one important loophole in this line of reasoning: the "defects" in space that might have been produced at the end of the GUTs epoch, which we will examine shortly.)

The other way to resolve the paradox is to propose some kind of dark matter as the culprit. If there happens to be some kind of dark particle that does not interact at all with radiation, then it can give a head start to the process of gravitational collapse. Dark matter would have been free to begin clumping together well *before* the moment when matter and radiation parted company at an age of about 100,000 years. Because radiation does not interact with these dark particles, it could not have interfered with their clumping process, as it did with baryonic matter before this time. Clumps of such dark matter could have achieved the necessary overdensities of better than 1 part in 1000 by the time of decoupling, but they would not lead to variations in the background radiation, thus evading that constraint. Such dark clumps would have served as seeds for eventual galaxy formation. The baryonic matter that glows and shines would subsequently collect upon these seeds like morning dew upon blades of grass, making them glisten with starshine.

Making galaxies in this way requires a lot of additional matter that is not composed of protons, neutrons, and electrons. It cannot be baryonic matter because baryonic matter could not begin to clump until the Universe was about 100,000 years old. And the amount of baryonic matter is severely limited by the Big-Bang nucleosynthesis arguments that we discussed in Chapter 4.

Appealing candidates for dark, nonbaryonic matter are the ghostly neutrinos, if they are indeed sufficiently massive. We know neutrinos exist, in great numbers, and that they parted company with baryonic matter and radiation when the Universe was about 1 second old. Such massive neutrinos would have had plenty of time to begin aggregating during the ensuing years. They could have begun forming dark swarms that were totally oblivious to the frenzied interactions of baryonic matter and radiation then going on all about. By the time this activity had subsided, at the moment of decoupling 100,000 years later, these swarms could easily have established the excess densities of 1 part in 1000 or better that are needed to induce subsequent gravitational collapse.

A problem with neutrinos, however, is that they prefer to clump only on the very largest distance scales, if at all. Because they are extremely light, they would have been very speedy and would have been able to form only very large aggregates corresponding today to the size of clusters, not individual galaxies. Much more ponderous, or slower, forms of dark matter are needed to make smaller, galaxy-sized clumps from initial overdensities.

Regardless of whether the dark-matter particles are slow or fast, however, overdensities in their distribution can begin to enter the linear growth phase only when their average density exceeds that of the radiation. For a universe with the critical density ($\Omega = 1$), this domination of (dark) matter over radiation would have occurred at an average temperature of 30,000°K. This point would have been reached at a time between the era of nucleosynthesis, which ended when the Universe was a few minutes old, and the moment of decoupling 100,000 years later — when the average temperature was 3000°K. In such a universe, we would have much more time available for linear growth of such aggregates of dark-matter particles. That means that primordial overdensities as small as 1 part in 10,000 could have spawned the galaxies and clusters seen today, assuming they began as clumps of dark-matter particles.

Such tiny overdensities in the primordial matter distribution would correspond to variations in the microwave background radiation of a few parts in 100,000. This is about the limit of present-day observations on the smoothness of this radiation. Hopes are high that, with the COBE satellite and the proposed experiments at the South Pole, these measurements can be improved in the near future — say to the level of a few parts in a million. If still no variations are seen at this level, it will be difficult to explain how today's galaxies and clusters could ever have formed from adiabatic fluctuations — even assuming the existence of dark matter.

As mentioned a bit earlier, there are other potential seeds for galaxy formation that are due to defects in the fabric of space, such as "cosmic strings" or "domain walls." Such topological entities may also be remnants of the GUTs epoch; in the late 1980s they became a subject of great theoretical speculation and excitement. They are hypothetical entities at present, but they represent very appealing ideas. In some sense, these defects behave like isothermal fluctuations, because they are independent of matter, but they have special, characteristic geometries.

Cosmic strings are extremely massive, one-dimensional strands and filaments that can form when the Universe goes from one phase to another, such as occurred at the end of the GUTs epoch. Should they exist, they would not be particles at all. They would be more like "cracks" in the very fabric of space itself. They would not interact with matter and radiation, except through the force of gravity. But cosmic strings would be so stupendously massive, with 1 inch weighing over a million billion tons, that they would make extremely effective seeds upon which matter could collect. Domain walls are similar defects, but they are two-dimensional sheets instead of one-dimensional filaments. We will return to these ideas again and again in the following chapters.

In order to make galaxies, we require two fundamental ingredients — matter and seeds. The matter can come in at least two varieties: luminous or dark. And most of the dark stuff is probably not baryonic. The seeds can be either random fluctuations in the primordial density of matter or topological defects in space itself.

———————————————— • ● • ————————————————

Whatever the case, we are faced with a key paradox that is telling us something important about the early Universe. On the one hand, matter today is clumped into some very dense, glowing regions able to foster and support intelligent life, which is able to peer back in time and wonder about its own origins. On the other hand, the dim afterglow of the primeval fireball shows that conditions were extremely smooth at the beginning, far smoother than we might have naively expected by looking at the structures visible today. Something new is needed — whether new kinds of matter, new physical theories, or new ways of thinking — if we are to resolve this paradox.

7

The Heavens at Large

*A*lthough enormous by earthly standards, the galaxy we inhabit is pretty insignificant when compared with the entire Universe. While light takes 60 thousand years to flash across the Milky Way, this is the briefest instant when we consider the 30 *billion* years it needs to traverse the portion of the Universe presently visible. The observable Universe is about a million times bigger than ordinary galaxies like our own. Indeed, if we could in some way compress this entire realm to the size of a huge cathedral — say Saint Peter's Basilica in Rome — then the Milky Way would be no larger than a tiny speck of dust drifting about inside.

We know, of course, that there are invisible halos made of dark matter extending well beyond the luminous fringes of visible galaxies. Perhaps these dark halos extend 10 times farther out, perhaps more. Even so, however, an entire galaxy including its surrounding halo would be no bigger than a large grain of sand on the floor of our cathedral. Were we able to play God and pack these concentrations of matter edge to edge, as tightly as possible, we could fit a million billion of them into the vast expanse of the Universe.

Galaxies are not, however, packed together in a random, haphazard fashion, like so many marbles in a bag. Nor do they

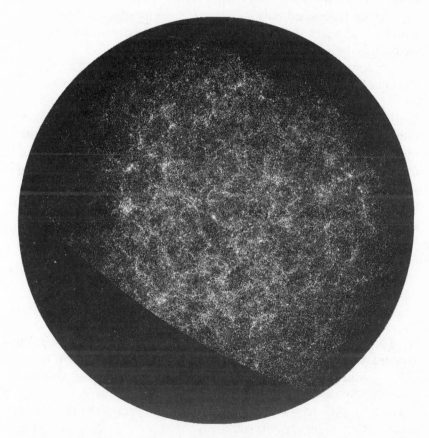

Distribution of galaxies as observed from Earth's northern hemisphere. In this view each galaxy is positioned at the point where it appears on an imaginary dome in the sky. This dome is centered upon the north galactic pole, which is located near the direction of the Coma cluster — the large, bright spot at the center of the circle.

resemble a disorganized swarm of dust particles floating aimlessly through space, independent of each other. The great majority of them seem to have congregated into large, orderly collections — the clusters and superclusters that we mentioned earlier. Looking in certain directions deep into space, we can find great ensembles of hundreds and even thousands of galaxies tightly packed into a relatively "small" corner of existence. In other regions there are vast empty holes, more than 100 million light-years across, containing hardly a single visible entity. The Universe appears to exhibit a definite *structure* even at such extremely large distance scales.

In recent years the detailed study of this large-scale structure has accelerated remarkably. We are just beginning to realize, for example, that galaxies are distributed throughout space in a pattern that resembles froth or foam. Clusters of them seem to congregate with one another more frequently than do the individual galaxies themselves. And some of these enormous clusters, including the one we inhabit, seem to be rushing headlong through space at tremendous speeds that cannot be due to the normal Hubble expansion alone. Such features, it turns out, may provide important hints about the nature of dark matter and the earliest moments of creation.

———————————————— • ● • ————————————————

In 1986, Valerie de Lapparent, Margaret Geller, and John Huchra of the Harvard – Smithsonian Center for Astrophysics (CfA) published a surprising result in the *Astrophysical Journal*. By determining redshifts for hundreds of galaxies inside a thin, pie-shaped slice in the general direction of the Coma cluster, they were able to establish the three-dimensional structure of the visible matter in that region to a depth of around 500 million light-years. According to their studies, galaxies are not sprinkled randomly about space. They seem instead to be arranged in regular patterns that look almost foam-like or sponge-like in appearance. The CfA astronomers found long filaments or thin sheets containing many galaxies surrounding vast empty regions with very few galaxies inside. It seemed as if the "slice of the Universe" under study had been cut right through a series of "bubbles," with the majority of galaxies residing on their surfaces.

Subsequent studies of adjacent slices revealed similar frothy patterns — as if one were peering through soapsuds. And one tremendous sheet of galaxies extends hundreds of millions of light-years across the entire field of view. Geller and Huchra dubbed this structure the "Great Wall." From these and other recent studies, it has become abundantly clear that galaxies are not randomly distributed at all. This is indeed an exciting development.

There had been earlier hints of such large-scale structure in 1978 when Robert Kirshner (then at the University of Michigan) and Gus Omler and Paul Schechter (then at Yale) made a red-

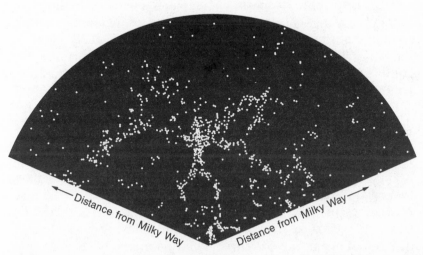

The first "slice of the Universe" surveyed by the Center for Astrophysics team. Each dot corresponds to a single galaxy, and the dense congregation near the center is the Coma cluster. Note the apparent filaments or sheets, plus the large voids with very few galaxies inside.

shift survey in a single direction looking toward the constellation Boötes. In such a survey, astronomers measure the redshifts of a sample of galaxies inside some well-defined portion of space. These redshifts are then converted into velocities of the galaxies relative to the Milky Way.

Once the redshift (or velocity) of a particular galaxy in their sample was known, Kirshner, Omler, and Schechter could estimate its distance from the Milky Way using the Hubble relationship between velocity and distance, assuming there were only small deviations from normal Hubble expansion. To their surprise, they found a large gap in their sample, stretching at least 200 million light-years across, in which there appeared to be hardly any galaxies at all. As one proceeded outward, the galaxies at first seemed to be bunched together, then there was this big gap, and then another bunch of galaxies.

Further observations in this same general direction have established that this "Hole in Boötes" (as it has come to be recognized) is indeed almost spherical. It is not just a long, skinny void that just happened to be aligned with our line of sight. It is a tremendous hole in space, encompassing over 10 trillion trillion cubic light-years, with very few galaxies inside. Even if we assume that the dark halo of an ordinary spiral galaxy (such as the Milky Way) extends 10 times as far as its luminous disk, we could still cram a *billion* of them into such an enormous volume.

Until 1986, skeptics could argue that astronomers had accidentally stumbled upon the one hole of any significance in the entire Universe. But the CfA surveys have silenced these critics. Instead of looking in only a single direction (called the "pencil-beam" approach), Geller, Huchra, and their colleagues have looked at all galaxies above a certain brightness in several adjacent slices of the Universe extending out about 500 million light-years. Because there are many more galaxies in a slice than in a pencil beam, the CfA surveys could not extend as deep into space as the earlier Boötes study. After converting these redshifts into distances, they concluded that galaxies were gathered mainly on large sheets or filaments—with huge voids opening up in between. As mentioned above, the large-scale structure of the Universe looks very much like soapsuds.

The CfA surveys have revealed huge voids almost as large as the one in Boötes in every portion of space large enough to contain one. Instead of their being rare occurrences, it appears that such voids are common features of the Universe. And the walls between the voids are remarkably thin—even though

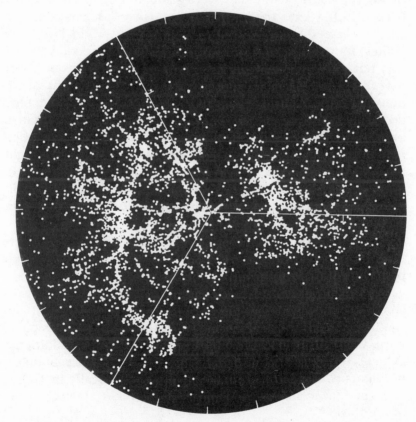

A 360-degree view showing the galaxies surrounding the Milky Way. The Great Wall is the huge structure at left, stretching at least 500 million light-years across the entire field of view in the north galactic hemisphere. At least 500 million light-years long, its size is so far limited only by the extent of the CfA surveys.

they stretch for great distances. The feature Geller and Huchra called the "Great Wall" extends at least 500 million light-years across their entire field of view. Its size seems to be limited only by the extent of their survey, and it could easily be larger still.

Another team of astronomers has peered even deeper into space and studied the large-scale structure *billions* of light-years away. David Koo of the University of California, Santa Cruz, and Richard Kron of the University of Chicago began their collaboration in the 1970s, while they were graduate students together at Berkeley; they have been continuing this work for more than a decade. Like the group that discovered the original Hole in Boötes, they measured redshifts along specific directions in the sky, but their surveys have stretched much farther out. They have repeated these measurements for many different directions, too, studying several independent pencil beams stretching deep into space. And Koo and Kron have teamed up with British astronomers Tom Broadhurst and Richard Ellis using a telescope in Australia so that their pencil beams could continue on right through the Earth to include parts of the southern sky, too. In effect, they are looking at a single "borehole through the Universe" centered on Earth and extending billions of light-years in either direction.

Together with Alex Szalay of Budapest, these astronomers found that galaxies are not sprinkled uniformly along the several billion light-years examined in each beam. Instead they are gathered together in distinct clumps separated by large gaps about 400 million light-years across in both the northern and southern skies. These early data seem to show a succession of great walls that are very regularly spaced. This is additional evidence that visible galaxies are distributed on sheets, bubbles, or filaments. The same kind of structure seen locally by Geller and Huchra seems to occur in the Universe at large. It is ubiquitous.

You may perhaps wonder why such striking cosmic features were not discovered much earlier, because astronomers have known how to measure redshifts for many years. The reason has as much to do with the sociology of astronomy as it does with

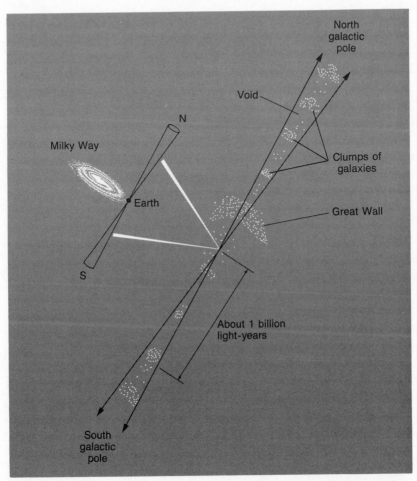

Two "pencil-beam" surveys made along the north and south galactic poles. According to these surveys, dense "walls" and sparse "voids" are commonplace features in the very large scale galaxy distribution.

the difficulty of measuring redshifts. While it is trivial to determine the position of any luminous object on a two-dimensional projection of the sky, to obtain the third dimension requires determining its velocity. To do that means measuring the Doppler shift of its light, as we mentioned in Chapter 1, and this

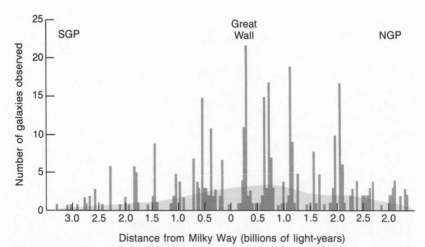

Distributions of galaxies observed in the combined pencil-beam surveys along the south galactic pole (SGP) and north galactic pole (NGP). The shaded area is the result expected if galaxies were distributed uniformly, which they clearly are not. Instead they seem to be clustered in thin walls that are spaced fairly regularly.

task requires a much longer time looking at the object. Astronomers must record this light until the individual atomic lines in its spectrum begin to stand out prominently from the background signals.

In the past, when astronomers have gone to such lengths to measure a redshift, they have usually studied objects deemed special — quasars, for example, or weird-looking galaxies called Seyfert galaxies. Or they have examined only a few galaxies near the center of a single cluster, such as the Coma or Virgo clusters, not the whole ensemble. The demand for observing time on the large telescopes needed to study redshifts has always been much greater than the supply.

In the recent measurements of large-scale structure, by contrast, astronomers have determined redshifts for all the visible galaxies in a designated sample. While such extensive surveys required painstaking efforts, and most of the galaxies studied were not particularly exciting in and of themselves, the overall result has been truly extraordinary. The Universe is structured like foam.

———————————— • ● • ————————————

While the astronomers mentioned above have been developing a geometrical model — a "picture" — of the large-scale structure of the Universe, others have been trying to describe it in mathematical terms. Led by James Peebles, groups of scientists have begun to *quantify* how galaxies and clusters are correlated with one other in space. How likely is one of these objects to be found in the vicinity of another?

During the 1970s, Peebles borrowed a helpful technique from solid-state physics and adapted it for use in studying large-scale cosmological structures. This method determines the excess probability — over and above what would be expected in a random distribution — that any two galaxies are separated by a given distance, which we will denote by the letter r. If we simply sprinkled galaxies randomly throughout space, there would always be some probability of finding pairs of them close together; what we really want to know is how much *more* than this minimum are they likely to be near one another. This *excess* probability is called the "two-point correlation function" of a system of galaxies.

For the Universe as a whole, this correlation function is inversely proportional to the 1.8th power of the distance between two galaxies; that is, it is proportional to $1/r^{1.8}$. If the separation doubles, the correlation function drops by a factor of about 3.5. For comparison, the force of gravity between two galaxies is inversely proportional to the square of their separation, or proportional to $1/r^2$. If we double the distance between them, the force of gravity falls by a factor of 4. Thus, gravity falls off slightly faster than the excess probability of two galaxies being a given distance apart.

Such a behavior seems fairly reasonable, because galaxies are thought to cluster with one another due to the influence of gravity. It is not at all hard to understand how gravitational clustering might induce such a power-law behavior of the correlation function. Because of gravity, it is more likely to find galaxies near one another rather than far apart. If we started with a random distribution of galaxies, gravity could induce such a clustering.

During the 1980s, groups in the United States and the Soviet Union began applying Peebles' technique to analyze how *clusters* of galaxies are correlated with each other. Neta Bahcall of the Space Science Telescope Institute and Raymond Soneira of Princeton, for example, studied clusters of varying "richness" —a measure of how many galaxies there are in a cluster's core. To their surprise, they found that clusters seem to be *more* strongly correlated than galaxies. If we look at one cluster, that is, we are very likely to find another cluster nearby. Bahcall and Soneira discovered that the cluster correlation function also seems to fall as the 1.8th power of the separation between clusters. Similar results were obtained independently by M. A. Klypin and Maxim I. Kolopov of the Space Research Institute in Moscow, and by Geller and Huchra in studies of the southern skies.

Some astronomers have questioned whether the apparent correlation between rich clusters might in fact be due to the effects of projections. A faraway cluster might seem close to a nearby (to us) cluster lying in the same general direction, or it might make the nearby cluster appear richer in galaxies than it actually is. Either effect would artificially bolster the correlation function for rich clusters. The final answer is not in yet on the effect of such projections, but there is good evidence for at least some enhancement in the cluster correlation function.

If clusters are indeed more strongly correlated than galaxies, it at first seems bizarre. And projection effects alone cannot explain the curious observation that the richer a cluster, the more strongly correlated it is with other rich clusters. When one considers that rich clusters are very rare objects in the Universe, and thus should be located extremely far apart, on the average, it seems absurd that they are more likely to be near one another than ordinary clusters and galaxies. What is going on?

If gravity were the only thing inducing the clusters to congregate, such an end result would be difficult to obtain. Imagine we started with a random distribution of objects that could subsequently move about only through the force of gravity. Then we would expect things that began the journey closer to one another (at greater density, that is) to be attracted more strongly and thus to be more highly correlated than objects that were originally much farther apart. On the very largest scales, gravity simply has not had a chance to do very much attracting, because gravity is extremely weak at large distances. But clusters, which are on the average much farther apart than galaxies, seem to be more strongly correlated.

If clusters are indeed more strongly correlated, something besides gravity is probably involved. And, in any case, we cannot escape the fact that large celestial objects like galaxies and clusters are *not* distributed at random. Gravity may not be the only influence determining the large-scale structure of the Universe.

———————— • ● • ————————

Additional evidence for large-scale cosmological behavior comes from examining how galaxies ebb and flow relative to the general cosmic expansion. Although the Universe as a whole is steadily expanding in the uniform manner first described by Hubble in 1929, there are deviations from this "Hubble flow," called "peculiar velocities." The Earth orbits the Sun at a distance of about 93 million miles; despite the Hubble expansion, this distance does not grow with time (fortunately). And the distance of the Sun from the center of the Milky Way does not vary much, either. In both these cases, the local force of gravity far exceeds the effects of the cosmological expansion, and the

Earth and Sun have peculiar velocities relative to the Hubble flow.

Similar effects are witnessed on even larger scales. The Milky Way, for example, is approaching the Andromeda galaxy, M31, not receding from it slowly as we would ordinarily expect due to the Hubble flow. Light from M31 reveals a blueshift — toward the blue end of the spectrum — not a redshift. And after we factor out the Hubble expansion, both the Milky Way and M31 seem to be falling toward the center of the Virgo cluster some 50 million light-years away. Peculiar velocities such as these can be attributed to the gravitational attraction between the various mass concentrations, which acts counter to the Hubble flow. Only at extremely large distances do we expect to find this uniform cosmological expansion becoming the dominant effect.

A peculiar large-scale motion was discovered in the late 1980s by a team of seven astronomers, dubbed "the Seven Samurai" (or "Seven Seekers of Truth," as they might prefer to be known), led by Alan Dressler of Mt. Wilson Observatory, Sandra Faber of Lick Observatory, and Donald Lynden-Bell of Cambridge University. By studying the light from a collection of elliptical galaxies, they showed that the entire Virgo cluster — including M31 and the Milky Way — is speeding in the general direction of the Hydra-Centaurus supercluster. And Hydra-Centaurus itself appears to be rushing ahead of the Virgo cluster in the same direction.

In fact, it seems as if a huge volume of space more than 100 million light-years across, including the Milky Way, the Virgo cluster, and Hydra-Centaurus, is moving coherently as a single entity and *not* spreading out with the Hubble expansion as one might expect. The peculiar velocity of this entire region is surprisingly large, about 600 kilometers per second. That's more than 1 million miles per hour! Compare this to the speed of the Solar System about the center of the Milky Way — around 230 kilometers per second, or half a million miles per hour.

This observation has been supported by further studies of galaxies in our general cosmic neighborhood. One group of astronomers (including John Huchra) examined a big ensemble

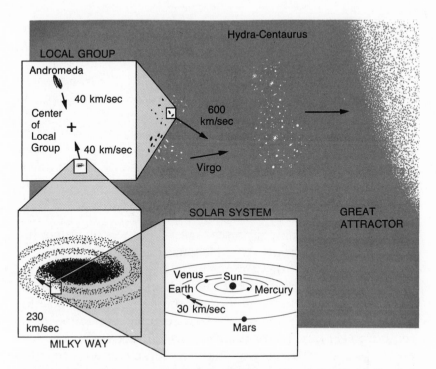

Peculiar motions in our corner of the Universe. The entire Local Group is moving together with the Virgo and Hydra-Centaurus superclusters at about 600 kilometers per second toward a region where a huge concentration of matter dubbed the Great Attractor is thought to exist.

of spiral galaxies instead of ellipticals, while the other group analyzed data gathered recently in the infrared portion of the electromagnetic spectrum by the IRAS (InfraRed Astronomy Satellite) experiment. Both groups found a pattern of peculiar velocities similar to that originally discovered by the Seven Samurai.

This large-scale motion agrees fairly well — in both its speed and direction — with what was obtained earlier from the dipole variation in the cosmic background radiation (see Chapter 6). Remember that this all-pervasive radiation is slightly redshifted in one direction and slightly blueshifted in the other, evidence that the Earth is moving relative to the rest of the Universe. Two totally different kinds of measurements seem to be telling us that our corner of creation is rushing headlong in the general direction of the Hydra-Centaurus supercluster. (To be a bit more precise, this direction lies between Hydra-Centaurus and the center of the Virgo cluster.)

Apparently the large-scale foam-like structure is not at all static like Styrofoam, nor is it expanding uniformly like sponge cake rising in the oven. Instead it seems to be churning to and fro like the foam on a storm-tossed sea. And the size of the underlying waves may be much bigger than a single cluster of galaxies. Each raisin in our raisin-bread model of the Universe, which should probably represent several clusters instead of just one, is thrashing about like a Mexican jumping bean!

We should be a little cautious here, however. An observation of a peculiar velocity is more difficult to make and harder to interpret than a measurement of the three-dimensional position of a galaxy. It requires not just an accurate evaluation of its redshift to get its velocity relative to us, but also a good estimate of the distance to the galaxy. Both measurements are needed to assess the *deviations* from the velocities one would normally expect in a completely uniform Hubble expansion.

As we discussed in Chapter 1, measuring distances accurately has always been difficult for astronomers — particularly at cosmological scales. One cannot just lay out a long ruler or use simple triangulation as a surveyor does, because the distances to other galaxies are tremendous. The Earth's motion about the Sun makes no discernible difference in the apparent position of another galaxy in the sky, so we cannot use such an effect to determine the distance to a galaxy. The Sun's motion about the Milky Way *would* make a difference, but who can afford to wait millions of years to witness the apparent change in position?

We therefore have to rely on "standard candles" in measuring such cosmological distances. These are objects whose intrinsic brightness we think we know a priori. We can then estimate the distance to them by comparing their apparent brightness in our telescopes with the assumed intrinsic value. The dimmer they appear, the farther away they must be. Unfortunately, finding reliable standard candles for galaxies is not all that straightforward. Therefore, large uncertainties always seem to creep into any determinations of peculiar velocities.

It is heartening, however, that precise measurements of the cosmic background radiation also suggest that the Milky Way is speeding in the general direction of Hydra-Centaurus. Fortunately, such a conclusion does *not* depend on any knowledge of cosmological distances. The smart money says that our corner of the Universe indeed has a large peculiar velocity relative to the Hubble flow.

• ● •

Attempts to explain this high peculiar velocity have focused on the possible existence of a huge concentration of matter lying beyond the Hydra-Centaurus supercluster at a distance of about 150 million light-years from the Milky Way. Dubbed the "Great Attractor," such a large aggregation — containing the equivalent of tens of thousands of galaxies, or more than a *quadrillion* times the mass of the Sun — could conceivably exert a net gravitational force on the galaxies and clusters in our own vicinity. If this concentration exists, it should pull them all relentlessly in the general direction of Hydra-Centaurus.

Efforts to locate the Great Attractor have already begun to reap rewards. Several groups of astronomers have found concentrations of rich clusters in the right direction, but around 500 million light-years away — too far to cause all of the observed effect on the galaxies in our nearby cosmic neighborhood. Other scientists have examined the distribution of light in our general vicinity, using existing galaxy catalogues or the IRAS survey and assuming that mass is spread about in the same way as the light. Both kinds of studies suggest that the peculiar velocities observed are due to the gravitational attraction of matter.

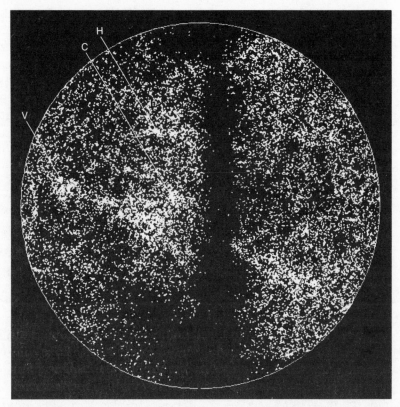

Plot of the galaxies in our vicinity. The Virgo (V) and Hydra-Centaurus (H and C) superclusters appear to be racing toward a Great Attractor, part of which may be the large aggregation to the right of Virgo and just below Centaurus.

Other teams of astronomers have been studying peculiar velocities at even larger distances, trying to determine the distance scale at which the Universe at last becomes completely smooth. If there is indeed such a Great Attractor, for example, the galaxies on the other side of it should have an overall peculiar motion in our direction. The Great Attractor would pull them *toward* us. Indications are that the flow does indeed reverse — at a distance of several hundred million light-years. In early

1990, Dressler and Faber presented evidence that galaxies on the other side are indeed falling inwards, as expected. Similar results were also obtained in another independent survey.

————————————— • ● • —————————————

The study of large-scale structure and motion represents one of the most exciting and fast-changing fields in cosmology today, with new and often surprising results turning up almost monthly. Whatever the lasting results of this effort, observes Alan Dressler of the Seven Samurai, "We have already established that even on a scale of hundreds of millions of light-years the Universe is made up of overdense regions, as well as regions relatively empty of matter." The tremendous Hole in Boötes, the Great Wall, and the large peculiar velocities of the Virgo and the Hydra-Centaurus superclusters are solid evidence for coordinated large-scale behavior even at such tremendous distances. How such large-scale features arose, given a very smooth initial distribution of matter, represents a major challenge for modern theories of the early Universe.

8

A Burst of Inflation

*A*lthough the Universe today appears extremely lumpy, with visible structures stretching across many millions of light-years, it clearly began as an extremely smooth and homogeneous medium. The nearly perfect uniformity of the cosmic microwave radiation (as discussed in Chapter 6) testifies to this inescapable fact. Much the same uniformity is seen today if we look only at the very largest distance scales — in the billions of light-years. A block of ice similarly appears homogeneous if observed at arm's length, but reveals tiny cracks, bubbles, and fissures when viewed up close.

It is natural therefore to wonder how the Universe ever became so smooth in the first place. Did it simply begin that way, or was there some sort of *process* that generated the overall uniformity from a state of primordial chaos? In either case, what caused the bumpiness we witness all around us today at more moderate distance scales? And how did the Universe ever become so old — about 15 billion years by recent estimates? These kinds of questions have intrigued cosmologists ever since the Big-Bang theory won broad acceptance.

Simple answers to them, and solutions to other pressing cosmological problems, can be found in a peculiar process called *inflation*, which arises in grand unified theories of the fundamental forces. This is not the same as the economic inflation government ministers and business executives try to cope with, but it does possess a few similarities. Cosmological inflation is an extraordinarily vigorous growth spurt that may have occurred in the first split second of existence — during the GUTs epoch, which would have ended at about 10^{-34} second. Such a stupendous explosion would have been far, far, far faster than the more leisurely Big-Bang expansion, whose effects we can still see today in the redshifts of galaxies.

In addition to solving several key cosmological problems in one grand sweep, inflation has another extremely important consequence. It requires that our Universe be an open universe with *exactly* the critical density — that is, with $\Omega = 1$. No other value of this parameter, which we discussed at length in earlier chapters, is possible if inflation occurred. And workable models of the inflation process seem to imply that the Universe we know and cherish may be only a single bubble of space, perhaps just one of many such bubbles drifting about, all completely out of touch with one another. Even though ours seems to be an enormous bubble, it may not be the only one in existence.

———————————— • ● • ————————————

In Chapter 6 we showed that the cosmic background radiation seems to be absolutely uniform, as best we can determine, coming with equal intensity from all directions and revealing the same effective temperature no matter where we look. The more one ponders the fact of this uniformity, the more unsettling it becomes. Remember that this radiation was emitted (or perhaps "released" is a better word) about 15 billion years ago, when the Universe was about 100,000 years old. The photons arriving now from two opposite directions, say north and south, began their long journey to our radio antennas in two regions of the Universe that today are almost 30 billion light-years apart due to the Hubble expansion.

Because the Universe is only 15 billion years old, there is absolutely no possibility these two regions could have transmitted any kind of signal from one to the other. Information cannot travel faster than the speed of light. Scientists say two such regions are "causally disconnected." What occurs in one should be completely independent of what is happening in the other. In fact, the cosmic background radiation coming from any two points in the sky separated by a few degrees of arc or more must have originated in two such distinct, causally disconnected regions. Light could never have traveled from one region to the other in the 15 or so billion years since the Big Bang. So how could they possibly have obtained enough information about one another to be at almost *exactly* the same temperature — to better than 1 part in 10,000?

If we think about the matter further, the mystery deepens. When this relic radiation was first emitted, the Universe was far more compact. The amount of mass within a given horizon (the distance light can travel since time began) was then but a tiny fraction of what it is today. The horizon distance was only about 100,000 light-years instead of about 15 billion. When the photons reaching two antennas today from two opposite directions were emitted, however, they must have originated at two points separated by 10 million light-years — almost *100* times as far as information could have traveled by that moment. This discrepancy is much worse than the one we mentioned above, which was only a factor of 2.

Our problem is to understand how any two such causally disconnected regions can possibly have one and the same temperature. Cosmologists have traditionally just assumed an arbitrary initial condition: the Universe simply *began* this smooth.

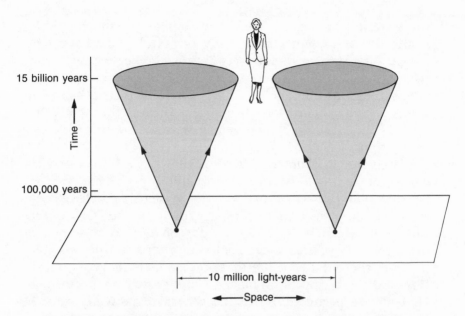

15 billion years

Time

100,000 years

|← 10 million light-years →|

← Space →

The "horizon problem." Observers of the cosmic background radiation re-
ceive signals from sources that were separated by 10 million light-years when
the Universe was about 100,000 years old. But the temperature of this radiation, coming from two such causally disconnected sources, is essentially the
same, as if each source somehow knows about the other.

Such a glib answer, however, merely begs the question. How did
this almost perfect smoothness arise in the first place? What we
would very much prefer is a concrete *mechanism* able to generate these conditions naturally, without our having to postulate
them. Physics, not metaphysics, should provide the answer.

A related mystery is the origin of structure. Although the
Universe is extraordinarily smooth at the largest distances, at
smaller scales it is extremely bumpy. Most of its visible matter is
found compressed into tiny pockets of great density — galaxies,
stars, and planets — relative to the universal average. How did
this graininess arise?

The same kinds of observations hold true for a body of water.
To the naked eye, water appears extremely uniform. Swirl your

hand around in it and you feel no bumps—only a smooth, continuous fluid. But a sufficiently powerful microscope will reveal water to be made up of countless numbers of individual H_2O molecules. And almost all the mass in these molecules is concentrated in tiny nuclei at the centers of their constituent atoms, which contain more than 99.9 percent of the total. What appears very smooth at first glance is found to be extremely grainy upon a closer look.

As we noted in Chapter 6, there are parts of the Universe today where the density is 30 orders of magnitude above the average. For such clumps to exist at all means that there had to have been some kind of primordial ripples or wrinkles in the fabric of the early Universe, which grew through gravity into clusters and galaxies, and eventually fragmented into stars and planets able to foster the intelligent life that now ponders its very origins. How did these ripples originate? There must have been some kind of tiny perturbations in the otherwise perfectly smooth initial distribution of matter and energy. Must we merely postulate these ripples? Or is there a way for them to develop naturally, from concrete physical principles?

———————————— • ● • ————————————

Although cosmologists may differ on the exact value, most agree that the Universe is around 15 billion years old, give or take 5 billion years. Its age is determined by a number of different methods. We can examine the age of the oldest stars in globular clusters, compare the relative amounts of radioactive substances found in meteorites, or estimate how much time it must have taken for galaxies to spread out to their present positions, based on the Hubble expansion rate. These three methods give ages between 10 and 20 billion years.

In standard Big-Bang cosmology, the density of the Universe evolves over time; it does not remain static. If this density is sufficiently high, as we discussed at length in Chapter 1, the Universe will eventually reach a maximum size and begin contracting; the average density will hit bottom, that is, and start rising again. If the density is low enough, on the other hand, the Universe will continue its expansion as the density keeps falling forever. In terms of the parameter Ω (omega), the ratio of the actual density to its critical value, contraction of the Universe can occur only if Ω is greater than 1, and its expansion to infinity happens only if it is less than or equal to 1.

The parameter Ω does not have to remain static, however. It can rise or fall too, because it is a ratio of two densities, the actual and the critical, both of which are evolving with time. If Ω happened to be greater than 1 at any moment, it would grow forever — approaching infinity as the Universe collapsed in a Big Crunch. In such a case, the actual density of the Universe (the numerator in the ratio) would be greater than the critical value; gravity would have enough strength to retard the steady outflow of matter, and the critical density (the denominator) would fall faster than the actual density. The ratio of the two densities, or Ω, would continue to grow ever larger.

If Ω was ever less than 1, on the other hand, that would mean that the actual density was less than the critical value at that moment. In such a case gravity does *not* have enough strength to retard the universal outflow, and the numerator of Ω falls faster than the denominator. Since it was less than 1 to begin with, Ω would fall steadily to 0 as the Universe expanded forever and ended in a Big Chill.

Only in the very special case where Ω is *exactly* equal to 1 does this parameter remain unchanged forever. Here the actual and critical densities are exactly matched, so they continue to evolve at exactly the same rate. In a critical-density universe, in other words, Ω must always be 1.00000000000. . . . If, for any imaginable reason, it was nudged higher by the slightest, tiniest amount, Ω would grow to infinity as the Universe subsequently collapsed. If it ever fell below 1, it would eventually drop to zero. The only way Ω can be 1 forever is to equal 1 exactly.

Similar arguments apply to a rocket ship blasting off from Earth. If it has exactly the right velocity (and direction) it will make it into a stable orbit. Just a bit higher velocity, and it can break free of the Earth's gravity and continue into deep space forever. If the rocket does not quite attain the necessary orbital velocity, however, it will soon crash back to the surface. And even if it does reach orbit, frictional forces due to collisions with stray gas molecules can eventually slow it down enough to bring it back to Earth in flames. This case corresponds to giving Ω a slight nudge above 1. If it ever becomes greater than 1, the Universe heads inexorably toward collapse.

Now if we take a closer look at the equations governing how the Universe evolves, we encounter a characteristic time for density changes to occur. This "dynamical time," which we can think of as the time required for the density to change by a factor of about 3, is inversely proportional to the square root of the density itself. The greater the density of the Universe, that is, the faster it evolves. If the density were quadrupled, for example, the dynamical time would be halved, and evolution would proceed twice as fast. If the density were divided by 100, the dynamical time would increase by a factor of 10, and things would happen much more slowly. The density evolves more rapidly in a denser universe, but the *speed* of this evolution increases only as the square root of the density.

Both the average density and the parameter Ω evolve on the same dynamical time scale, which in round numbers is about equal to the age of the Universe at any given moment. The density of the Universe at present, for example, is changing on a scale of 10 to 20 billion years, which is roughly how long it will take for this density to fall by a factor of 3 from its current level. In the very early Universe, when it emerged from a spacetime foam with an absolutely stupendous density, the dynamical time was exceedingly brief — 10^{-43} second (or less than one-millionth of a trillionth of a trillionth of a trillionth of a second). At that very instant, the density (and Ω) was changing extremely rapidly, taking only 10^{-43} second to fall by a factor of 3. Even by the end of the GUTs epoch a bit later, the moment of matter creation, such a change would have taken only 10^{-34} second. By the era of nucleosynthesis, where we have the abundance of light elements to verify our understanding of this evolution, Ω was changing on a time scale measured in seconds.

In its earliest stages, therefore, the Universe was evolving extremely rapidly. Had it ever been slightly greater than or less than 1 at any moment during those times, Ω would have zoomed to infinity or plummeted to zero in a fraction of an eyeblink. Only in the special case where Ω equaled 1 *exactly* could it have ever remained constant and not have changed with time. Unity, it seems, is a very unstable equilibrium point. Go even the slightest bit off 1, and Ω rises or falls with absolutely blinding speed.

As mentioned above, the Universe is about 15 billion years old, which is over 60 orders of magnitude greater than 10^{-43} second. If the Universe had emerged from the spacetime foam with Ω differing from 1 by even the tiniest amount, say in the *sixtieth* decimal place, this parameter would have evolved significantly by now and be well on its way to zero or infinity. At present, although we cannot say for sure that $\Omega = 1$ exactly, we do know that it is greater than 0.1 and less than 3 — or within an order of magnitude of 1. To be so close today, Ω must have been extremely near 1 during the earliest instant (at 10^{-43} second). Any possible difference had to be near the sixtieth decimal place or beyond.

How did such a fine tuning of Ω, to an initial value so extraordinarily close to 1, ever occur? Or, equivalently, how did the Universe become so incredibly *old*, without ever having evolved to an infinite or zero density? This mystery is known to cosmologists as the "age problem." And as $\Omega = 1$ corresponds to a Universe in which a flat, Euclidean geometry is the norm (see Chapter 1), it is also referred to as the "flatness problem." It is another important problem of initial conditions that we would prefer to solve using physical principles.

• ● •

Another problem that arises in grand unified theories of the early Universe is known as the "monopole problem." We mentioned this before, in Chapters 1 and 5, where we noted that massive particles known as magnetic monopoles should have been created in abundance when the symmetry of the GUTs force evaporated — as it broke down into separate, disparate forces. In the late 1970s, the Soviet theorist Sasha Polyakov (now at Princeton) proved that creation of these exotic particles *must* occur whenever any unified force breaks down to the electromagnetic force (as well as other forces). Just as the GUTs breakdown was producing baryons and generating a tiny excess of matter, therefore, it should also have created a comparable number of these ultraheavy particles, over a million billion times more massive than a proton, which possess the magnetic characteristics of a single north or south pole.

Although there have been occasional reports claiming to have found one, magnetic monopoles have never been convincingly detected. The few "discoveries" have never been confirmed or duplicated. Even if these reports *were* true, it has become absolutely clear that there are far fewer magnetic monopoles than baryons in the Universe. And thankfully so, too. Otherwise they would have easily dominated the mass of the Universe and triggered its collapse long ago. You would not be here to read these words. (Nor would we have been around to write them!) Somehow the Universe must have found a way out — a means of getting rid of all its monopoles, or at least the great majority, and still keeping its baryons intact.

One last quandary that cosmologists wrestle with is the question, "Why is the Universe not rotating?" Almost every celestial object — suns, moons, planets, and galaxies — rotates in one way or another. Why not the entire Universe itself?

To speak about any rotation, in fact, one must specify an axis about which the circular, rotary motion occurs. But the Universe has no center, it has no preferred point or axis. All points in it are equivalent. How, then, did the Universe attain this special state — with all points equivalent and no axis of rotation?

———————————— • ● • ————————————

Prior to 1980, finding a solution to these pressing cosmological problems — smoothness, bumpiness, age, flatness, monopole, and rotation — required us to set the initial conditions of the Big Bang arbitrarily. These conditions could not be justified a priori; they had to be postulated. But in that year Alan Guth, who (as we mentioned at the beginning of this book) was then a postdoc at the Stanford Linear Accelerator Center (and is now professor of physics at MIT), published a dramatic paper outlining a way to solve all these problems simultaneously. In the decade since his great insight, a completely new picture of the early Universe has emerged.

The previous fall, Guth had been working with Henry Tye of Cornell, trying to solve the monopole problem, which seemed to raise its ugly head whenever grand unified theories were applied to the early Universe. Working at home late on the evening of December 6, 1979, he found a way to solve this problem that also explained why the Universe today is so old and flat. "I discovered that the Universe grew exponentially, inflating like a balloon," he recalled. "I felt it was a spectacular realization."

Guth dubbed his breakthrough idea "inflation," partly because of the terrific explosion that would have rocked the early Universe. And monetary inflation had been racing through national economies during the late 1970s, too, so it seemed an obvious name choice. Cosmological inflation is not an alternative to the Big Bang. It is a unique and powerful way to set the initial conditions of the Big Bang during the GUTs epoch, at a time of about 10^{-34} second.

To understand how this revolutionary process of inflation works, remember that the expansion rate of the Universe is determined by its total energy density — the sum of matter and radiation per unit volume — at any instant. The greater the energy density, the faster the Universe will expand. Because the early Universe was extremely hot and dense, almost beyond our comprehension, it was expanding extremely rapidly. But as time elapsed and this expansion continued, as more space was added to the Universe but no more matter or radiation, the energy density necessarily dropped, and with it the rate of expansion. The existing energy total had to be distributed throughout an ever-growing volume, lowering the universal energy density and slowing the overall outrush.

Suppose there were a way, however, to hold the energy density *fixed* even though space itself kept expanding. Then we would attain a truly inflationary condition wherein the expansion rate far exceeded the more leisurely pace of the normal Big-Bang expansion. The strong outward push would continue unabated until this condition ceased to hold true. The Universe would quite literally *explode*.

But how, you may well ask, can we hold the energy density fixed while space itself is in the process of exploding? Doesn't this stupendous outrush immediately dilute the total energy inside, spreading it around more and more sparsely? That would seem to be the natural consequence. But if the new space being added *also* contributes energy as it comes into being, then the total energy density can indeed remain relatively constant, and inflation would continue relentlessly.

This scenario may appear paradoxical, given our strong prejudices against creating matter and energy from essentially nothing at all. But by giving space itself an intrinsic amount of energy we do something similar to the effect of the "cosmological constant" in Einstein's theory of general relativity. This parameter reflects the overall curvature of space itself. General relativity tells us that the curvature of space is closely related to gravity and thus to the energy density; space is wrinkled in the presence of matter and energy. If "empty" space also possessed an energy density, it would mean that space itself has an intrinsic curvature even in the complete absence of matter. As Einstein showed explicitly, such a "vacuum curvature" can drive the expansion of the Universe.

If somehow during the early Universe we could have reached such a condition whereby space itself possessed energy all its own — where the vacuum, that is, had a net positive energy — this vacuum energy could have driven a burst of inflation. It may seem nonsensical, that the void might have contained something intrinsic to it, because the vacuum is generally thought to be the *absence* of everything, but grand unified theories raise just this possibility.

In unified theories with a property called "spontaneous symmetry breaking," there is a unique feature called the "Higgs field" (after theorist Peter Higgs of the University of Edinburgh, who first conceived the idea in 1964). This field permeates existence and can effectively give the vacuum a positive energy density. The symmetry breaking, which is thought to have occurred through the agency of this Higgs field, took the Universe from a condition where there was only a *single* force at high energy (or temperature) to one where there were several different forces at low energy — the way things are today.

When the unified electroweak force broke down into the weak and electromagnetic forces, at about 1 picosecond (10^{-12} second) into creation, such a Higgs field was responsible for the seeming disparity between them. This field is what causes the masses of particles, including the particles that carry forces. It gives the W and Z bosons (which bear the weak force) masses of 80 and 91 GeV, while the photon mediating the electromagnetic force remains absolutely massless. Thus, the weak force appears extremely feeble, because it is difficult to produce such massive carriers, which can therefore act only over very short distances, far smaller than the size of a proton. It is easy to produce massless photons, on the other hand, and the effects of electromagnetic force can extend to great distances.

A similar series of events may have occurred at about 10^{-34} second, when a GUTs force broke down spontaneously into the strong and electroweak forces. When the initial symmetry was shattered, the vast energy of another Higgs field, say an "ultra-Higgs" field, became bound up in the masses of the force-carrying particles — here the absolutely gargantuan X and Y bosons, with masses around 10^{14} GeV. At this point the energy of the vacuum itself fell essentially to zero.

Just before that moment, the energy of the ultra-Higgs field inhabited space itself, so the energy density of the vacuum was not zero then, but had a finite (and extremely large) value.

Today we know the energy density of the vacuum to be zero, or at least very close to it. But as the early Universe passed from a condition where all the forces were unified and a perfect symmetry reigned supreme to its present state of broken symmetry, the vacuum itself contained energy, which is what broke the symmetry. Lots of energy—something like 10^{95} ergs per cubic centimeter. In more familiar units, that's about 10^{84} kilowatt-hours in every cubic foot. For comparison, the energy density of water (using $E = mc^2$) is 10^{21} and that of atomic nuclei is 10^{36} ergs per cubic centimeter. Thus, the energy density of the vacuum just before 10^{-34} second was almost a trillion trillion trillion trillion trillion times that of atomic nuclei today!

At the extremely high temperatures of the very earliest Universe, well before 10^{-34} second, the total energy density was much greater than even this stupendous value. So the progress of expansion before that moment would have followed the standard recipe, with its pace falling gradually as space unfolded and the energy density decreased.

But once the Universe had cooled to a temperature where the ultra-Higgs field appeared, it would have generated a vacuum energy that began to dominate the total energy density, and things must have begun to accelerate dramatically. Now as additional space unfolded, more vacuum energy was produced too, keeping the overall energy density of the Universe relatively constant. And since the expansion rate is proportional to the energy density, the Universe continued to expand at the same fixed rate, adding still more space and driving the expansion even further.

Such an inflationary period would have continued until the Universe passed through a transition at the end of the GUTs epoch, when the symmetry between forces finally broke down. At this moment the ultra-Higgs field gave the X and Y particles effective masses, which necessarily broke the symmetry between forces. As these particles acquired masses, the energy of the vacuum fell essentially to zero. Radiation again began to dominate the total energy density, which drove the far slower Big-Bang expansion observed today. The pent-up energy of this ultra-Higgs field became manifest as an expanding fireball of matter and light, in which the perfect symmetry of the original unified force had been broken forever.

During this inflationary period, the Universe literally exploded. Its size grew exponentially from the moment the vacuum energy density began to dominate the total until the phase transition ended this fantastic spree at the close of the GUTs epoch. (When something grows exponentially, it is multiplied by a given factor, say 10, in each interval of time. Thus, it might be 1 in the first interval, 10 in the next, then 100, 1000, and so on. The numbers get extremely large very fast.) By the end of this era at about 10^{-32} second, the distance between two points would have swollen by at least 50 orders of magnitude — or by a factor of 10^{50} — as a result of inflation. Things would never be quite the same again.

• ● •

Using the concept of inflation, we can now resolve the problems we discussed earlier. The smoothness of the observable Universe is not such a perplexing mystery any more because the region from which it originated (before 10^{-34} second) was a very, very tiny realm where everything was completely in touch with everything else. In fact, it was less than 10^{-34} light-second (or about 10^{-24} centimeter) across before inflation, the distance light travels in 10^{-34} second. That's about a trillionth the diameter of a proton!

During inflation that tiny region swelled *tremendously*, to a domain at least the size of a grapefruit and perhaps even more than a billion billion kilometers across. All the matter we can see in the Universe today had to originate inside such a tiny region, where everything would have been in thermal equilibrium before inflation. Thus, it is easy to understand how the regions of the Universe that appear to be causally disconnected today can

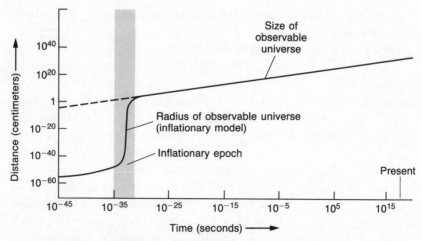

Expansion of the Universe before, during, and after inflation. The size of what is today the observable Universe swelled by a stupendous factor of at least 10^{50} during the brief inflationary epoch.

be at exactly the same temperature. They started out that way, long, long ago — before inflation occurred.

Inflation also solves the age, or flatness, problem — why the Universe is so old or flat. It turns out that the terms in the equation that can cause Ω to differ from 1 are inversely proportional to the square of the characteristic size of the Universe, its so-called "scale factor" R. If R doubles, that is, these terms become smaller by a factor of 4; if R increases by a factor of 10, they decrease a hundredfold. But during inflation the scale factor of the Universe swells by many, many orders of magnitude, and these terms become absolutely negligible — no matter what their initial value might have been. Thus, inflation drives Ω inexorably toward 1.

We consequently live in a very old Universe where the flat, Euclidean geometry holds sway. If Ω were not equal to 1, as we discussed in Chapter 1, then space would be curved — having either a spherical ($\Omega > 1$) or hyperbolic ($\Omega < 1$) curvature. Let's assume things started out either way before 10^{-34} second.

What inflation does is force the scale of any preexisting curvature (related to R) to be so immense that the Universe has to appear completely flat afterward. The radius of the Earth is large compared with Illinois, for example, so Illinois appears flat when in fact it has a small curvature.

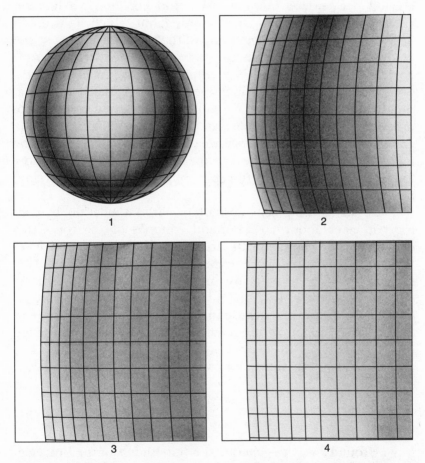

How inflation yields a flat Universe. In each drawing the original sphere is inflated by a factor of 3; the curvature of the sphere quickly becomes virtually undetectable.

Inflation eliminates the rotation problem in a similar fashion. The terms describing the rotation of the Universe about any axis are inversely proportional to some power of R. Therefore, they vanish completely during inflation, as R itself virtually explodes, and we end up with *no* apparent rotation today.

The monopole problem is solved, too, because only a few monopoles can be created in an inflationary universe, and we can never expect to encounter them. Only one monopole can be created per horizon volume just before inflation. (A horizon volume is the volume of space throughout which light could have traveled since the beginning of the Universe.) Because the entire observable Universe must have originated from a *single* horizon volume about 10^{-24} centimeter across just before inflation, that means only one monopole in all. Actually, the situation is more complex, and we might expect a few more monopoles because of events that can occur after inflation, but the grand total is still very, very small. Unless we happen to be incredibly lucky, we can never expect to see one, no matter how long we search. Some intrepid physicists are still trying, however!

In the standard Big-Bang cosmology without inflation, by contrast, it takes about 10^{80} horizon volumes during the GUTs epoch to create enough matter to make up our present Universe. Similarly, we would have expected to generate about 10^{80} monopoles—about as many monopoles as baryons! This is clearly not the case. The total number of baryons is not affected by inflation, however, because they were created at the very end of the GUTs epoch, *after* inflation ceased. We still produce plenty of baryons in an inflationary Universe, giving us firm ground to stand on and lots of sunlight to brighten our days.

———————————— • ● • ————————————

Having found a way to generate the smoothness of the Universe, we still need to explain the obvious bumpiness—the galaxies, clusters, and other large-scale structures. This turns out to be a knotty problem that has required further elaboration of inflation beyond what Guth had originally conceived.

Some bumpiness must have emerged at the close of the GUTs epoch, when the symmetry of the grand-unified "superforce" was broken, inflation ceased, and matter was created. At this moment the Universe went through what is called a "phase transition," akin to the process that transpires when steam condenses into liquid water, or water freezes into ice. These are just the three different "phases" of H_2O — and condensation, freezing, boiling, and melting are transitions between them. Phases that occur at higher temperatures possess a greater internal symmetry. The gaseous phase has greater symmetry than the liquid, which itself is more symmetric than the solid phase. As things cool down, individual H_2O molecules that once were free to roam about become locked into more rigid, asymmetric patterns. The entire collection of molecules experiences a phase transition at the same time as its symmetry breaks down.

At the end of the GUTs epoch, the Universe would have passed through a similar transition from a symmetric phase, in which the strong, weak, and electromagnetic forces were completely equivalent, into an asymmetric phase — in which they are not. Although still the very same Universe, it now appears decidedly different from what it was before 10^{-34} second.

How did the Universe get from one phase to the other? As it cooled, the original, perfect symmetry broke down whenever the temperature fell below a certain level — about 10^{27} (or 1 billion billion billion) degrees. Small regions of asymmetric space must have formed wherever this condition was met. For the phase transition to be complete, however, each region somehow had to coalesce with all the others. This was not a trivial feat.

Imagine you are trying to freeze a certain volume of water, say a few liters. Tiny ice crystals begin to form and grow at random, slowly merging with one another to form a multicrystalline block of ice. But what would happen if the water happened to be flying apart faster than the crystals ever could grow? The water could never freeze into a single block of ice.

There was a key assumption being made above, namely that you were doing this freezing experiment on the Earth's surface, where atmospheric pressure and the force of gravity combine to keep the water confined in whatever container is being used. If you tried to do it in outer space, the water molecules would fly apart immediately, and no freezing could occur.

Perhaps a better analogy is what happens in a snow machine, which is used to tailor ski slopes when nature fails to do the job itself. Water is sprayed at high speed out of a nozzle and freezes almost immediately upon contact with the cold night air. But the individual droplets are moving apart too fast to coalesce. The ice crystals in different droplets, the granules of the new phase, can never run into one another. We only get individual grains of snow, each with its own crystal structure, not a block of ice.

Cosmologists often use the term "bubbles" to refer to the individual grains of asymmetric space formed during the GUTs phase transition in the early Universe. Here they are invoking an analogy with boiling water (which, unfortunately, is a transition from a lower to a higher symmetry phase, the opposite of what occurred during the GUTs phase transition). When water boils, separate bubbles emerge in the liquid and subsequently run together, eventually converting all the available liquid into steam. When the Universe finished the GUTs transition, the individual "bubbles" of its final phase should have merged.

One might expect to comprehend how the bumpiness of the Universe arose by understanding how its grains, or bubbles, of space coalesced as the GUTs transition ended. Consider again what happens in a static body of water when it freezes. Small irregularities occur as the many different crystals merge with one another. In an inflationary scenario, however, there is an important hitch: grains of the new phase — its individual bubbles — can be moving apart so fast that they will *never* merge. The GUTs phase transition would never be completed. This, in fact, is essentially what happened in the original theory of inflation advocated by Guth.

The solution to this difficulty was proposed in 1982 by Andrei Linde of the Lebedev Physical Institute in Moscow — and independently by Andreas Albrecht and Paul Steinhardt of the University of Pennsylvania. The actual GUTs phase transition, they suggested, might have been a very smooth kind of transition, of a type originally studied by Erick Weinberg at Columbia and Sidney Coleman of Harvard. According to this reasoning, the entire observable Universe would have originated inside a *single* bubble (or grain); individual bubbles would not have needed to overtake each other in order for the phase transition to end. And by requiring that the transition be extremely smooth, with the total energy distributed uniformly throughout the bubble, these scientists could ensure that the Universe would now seem homogeneous — uniform in all directions. Called "new inflation," the method also allowed the production of enough matter at the end of the GUTs phase to be consistent with present observations.

The idea that the Universe arose from a single bubble during the GUTs transition has tremendous implications. Each horizon volume that existed before inflation could have evolved into its very own independent universe. This means that there were probably many, many other bubbles at the GUTs phase transition — all of them leading to other, parallel universes totally cut off from our own. The science fiction writers will probably have a field day with this idea.

Although there may indeed be many other universes "out there" that we never before contemplated, we will have no way of traveling to them or knowing about them. Even the starship Enterprise of "Star Trek," with its warp drives that can propel its intrepid heroes from one galaxy to the next at amazing, faster-than-light speeds, would be of little use in hopping from one universe to another. The "space" between bubble universes is not our familiar empty space at all, but an undifferentiated, symmetric space where all forces are still unified in a superforce. It is a realm where all particles are identical and normal matter does not exist. There we would encounter only pure

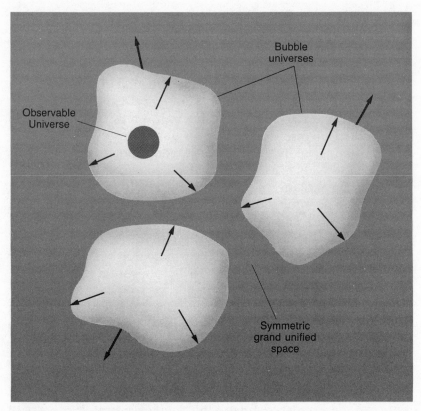

Bubble universes that might occur in a theory of new inflation. Our own, observable Universe may be just a small part of one such bubble universe, which is completely cut off from many others.

Higgs energy—the tremendous energy of the ultra-Higgs field —which would effectively be everything possible, all things at once.

Despite its successes, new inflation had difficulty generating enough bumpiness to accord with what is observed in the Universe today. If the entire Universe began as a single, smooth, solitary bubble, then that bubble somehow had to generate all primordial ripples (or cosmic seeds) *within* itself. We can no longer count on collisions between individual bubbles to do the trick. And because of inflation, we cannot rely on any ripples that might have existed earlier, before the GUTs epoch. These would have been wiped out completely during the extremely rapid expansion due to inflation, like the wrinkles on the surface of a balloon as it is blown up. No, any ripples that could have grown eventually into clusters and galaxies must have begun inside the one bubble itself, during or after inflation, if the theory of new inflation holds true.

Several physicists including Guth and Steinhardt began to attack this flaw during the summer of 1982, at the Nuffield Workshop on the Very Early Universe. Soon an interesting solution emerged. In a single bubble, it seems, there would have been random oscillations due to what are called "quantum fluctuations" of the ultra-Higgs field. The space being created during this phase transition would have "quivered" slightly, like a great mass of gelatin, because of the famous Heisenberg uncertainty principle. According to this principle, physical variables such as energy, time, position, and momentum can never be known exactly. In the microscopic quantum domain, there is always *some* irreducible uncertainty. The measured values of such variables can fluctuate within certain well-defined limits.

In the early Universe, such quantum fluctuations could have produced small perturbations in the otherwise uniform density of a single bubble. Inflation would then have blown these tiny ripples up to macroscopic sizes. Eons later, they would have triggered the gravitational collapse that led to the formation of all galaxies and clusters. From the very tiniest, most insignificant wrinkles on the fabric of spacetime, we eventually obtain the largest structures in all of existence.

In fact, these quantum fluctuations possess a very desirable feature. They have just the mass distribution proposed during the early 1970s by Edward Harrison and Yacov Zeldovich (see Chapter 6). In order to explain the observed masses of galaxies and clusters, these two astrophysicists had independently suggested that the probability of primordial density ripples (with a given mass) must have dropped slowly as the mass increased. Thus grand unified theories, through the mechanism of new inflation, can produce the required mass distribution automatically. No arbitrary assumptions are needed.

Although the mass distribution comes out right, the amplitude (or size) of these primordial ripples does not. The simplest grand unified theories lead to large overdensities that would have begun their collapse far too early in the history of the Universe. From Chapter 6 you may remember that any overdensity must have been less than 1 part in 1000 at the moment radiation decoupled from matter, else there would be very obvious variations in the cosmic background radiation. In the simplest GUTs (such as Georgi and Glashow's SU5 theory), quantum fluctuations from the single-bubble phase would have been 50 parts in 1, not 1 in 1000. Such a tremendous overdensity would have already collapsed into a black hole long before. Galaxies could never have formed.

Cosmologists and particle theorists soon began searching for grand unified theories that did not possess such a serious flaw. The requirement that these primordial ripples not be too large is an important constraint on viable theories. Although the final word is not yet in, GUTs that incorporate the feature of supersymmetry, called "SUSY-GUTs," are the apparent favorites. As discussed in Chapter 5, supersymmetry may be necessary to unify the GUTs force with gravity, and it predicts a whole new class of particles that may be one source of dark matter. SUSY-GUTs can limit ripple sizes while providing almost everything else — inflation, bubble universes, and the correct mass spectrum. Their advocates think such theories may truly be the key to the Universe.

— • ● • —

All theories of inflation make a dramatic prediction by which they can be readily tested. In the process of solving the age or flatness problem, inflation requires that Ω be almost *exactly* equal to 1. No other physical theory has been able to specify its value so precisely and unconditionally. In the past, cosmologists had to rely primarily on guesswork or intuition. Now there is a specific, testable prediction.

The requirement that $\Omega = 1$ has another interesting consequence. Because Big-Bang nucleosynthesis arguments (see Chapter 4) allow baryons to contribute at most 10 percent of the matter needed to close the Universe, there must be a lot of nonbaryonic matter around, if inflation is true. In fact, the great bulk of matter—between 90 and 95 percent—would have to be nonbaryonic. And since the dark matter that clumps with galaxies and clusters cannot account for Ω greater than 0.3, as discussed in Chapter 3, we would have to conclude that most of the nonbaryonic dark matter is not associated with a visible object. If inflation is true, the bulk of the dark matter is completely unlike normal, garden-variety matter made from quarks, and it does not even cluster with the stuff we can see.

How, then, does the inflationary prediction for Ω stack up against observations? From examining the dark halos of galaxies and clusters, we know that Ω is probably greater than 0.1 and perhaps as large as 0.3. Thus, the lower limit on Ω is within a factor of 10 of the value predicted by inflation. By cosmological standards, that's pretty close.

An upper limit on Ω can be obtained by studying the Hubble "constant" as we look backward in time—deeper and deeper into space. From Chapter 1, you may remember that the Hubble constant equals the recession velocity of a galaxy divided by its distance from the Milky Way; this number gauges the rate of universal expansion. When we look at a galaxy that is billions of light-years away, however, the ratio of its velocity to its distance gives us the Hubble constant as it was billions of years ago, not today. As we peer back to these earlier times, therefore, we expect this ratio to be *greater* than it is in our local neighborhood today, because the expansion of the Universe has been slowing down over time.

Comparing yesterday's values of the Hubble constant with today's gives us the "deceleration rate" of the Universe. Although it is hard to measure this slowdown accurately, due to difficulties in determining such extremely large distances, it does indeed seem true that the deceleration rate is not overly large. Otherwise, there would be big and easily measurable deviations from a constant Hubble expansion, and these are not seen.

In the standard Big-Bang cosmology, the deceleration rate is directly related to the density of the Universe, and thus to Ω. The greater the average density, the greater the gravitational pull of the Universe upon itself, and the faster the Hubble expansion slows down. From the most conservative limits on the deceleration rate, we can conclude that Ω is less than about 3. This is very close to the inflationary prediction, by cosmological standards.

Princeton physicists Edwin Loh and Earl Spillar have attempted to put more stringent limits on the value of Ω, using the fact that a high-density (or high-Ω) universe should have more galaxy clusters per unit volume than a low-density universe. Using a 90-inch telescope in Wyoming, they made redshift surveys of over a thousand galaxies in five patches of the sky. From the measured density of clusters at larger and larger distances, they estimated Ω to be about 0.9, with an upper limit of 1.6 and a lower limit of 0.4. Taken at face value, this result is a strong confirmation of the inflationary prediction and proof of the need for nonbaryonic dark matter in the Universe.

But many astronomers have argued that the measurement techniques of Loh and Spillar involve some debatable assumptions—and that their data are telling us more about the evolution of galaxies over time than about the value of Ω. These criticisms, however, may reflect professional jealousy, at least in part. Loh and Spillar are physicists, not astronomers, and the old guard may be a bit too zealous in its criticism of these young upstarts. Obviously the final word has not been uttered on their approach. Further observations are needed.

Another way to measure Ω is to use the peculiar velocities of galaxies to get an idea of the underlying mass distribution. As you recall from Chapter 7, the relatively high values of these velocities in our own galactic neighborhood provided the initial evidence for the Great Attractor. By examining such peculiar velocities over huge volumes hundreds of millions of light-years across, astronomers can determine the distribution of massive objects inside that must be causing these motions. Early results of two such studies, one led by Edmund Bertschinger of MIT and the other by Nicholas Kaiser of the Canadian Institute of Theoretical Astrophysics, both indicate that Ω must be close to 1 if we are to explain the peculiar velocities observed. On a scale of hundreds of millions of light-years, they claim, Ω must be greater than 0.4. If these results hold up, they will verify the need for nonbaryonic dark matter.

Similar results were obtained earlier in a study of galaxies observed at infrared wavelengths by the IRAS satellite. Teams led by Marc Davis of Berkeley, Amos Yahil of the State University of New York at Stony Brook, and Michael Rowan-Robinson of Queen Mary College in London mapped the velocities of these infrared galaxies out to several hundred million light-years. The distribution of these velocities also seemed to require that Ω be close to 1.

These kinds of measurements are still only in their early stages. In the near future we can expect that they will be improved and refined — and their ambiguities resolved. Whatever the final conclusions, however, it is clear that Ω is close to 1, at least by cosmological standards. Convincing proof that the Universe has passed through an inflationary epoch may be almost at hand.

---------------------- • ● • ----------------------

In the decade since it was first proposed, the theory of inflation has become one of the dominant ideas in all of Big-Bang cosmology. Though still a hypothesis, it is a very compelling one that many cosmologists already take for granted. It has become increasingly difficult to imagine how our Universe could have begun without invoking something like inflation to clear away the monopoles and make it end up so flat, no matter what it was like before. And experimental proof of inflation — that Ω is very close to 1 — seems to be almost at hand. If and when it comes, we will have to confront the likelihood that over 90 percent of the Universe is totally dark and not made of ordinary matter.

9

The Shadow World

Wherever we gaze in the Universe, it appears, there is far more than the eye can possibly see. The motions of galaxies and clusters, the rapid formation of galaxies, and the theory of inflation all require the existence of dark matter. Without it we are at a loss to explain these phenomena, and we have great difficulty understanding why the Universe is so flat, smooth, and monopole-free. But even though the evidence for *some* variety of dark matter has become fairly convincing in recent years, the exact nature of this pervasive cosmic ectoplasm still remains a mystery. We turn now to this important question.

The existence of huge dark halos surrounding the luminous regions of galaxies depends on very few assumptions, as we noted in Chapter 3. Either Newton's universal law of gravity is not as universal as originally imagined, or there are vast shrouds of unseen matter extending far beyond the visible core of virtually every galaxy. No other options exist. What this ghostly stuff might be we cannot say for certain, but we can make some educated guesses.

From the arguments about Big-Bang nucleosynthesis, as discussed in Chapter 4, we know that there are not enough baryons around to close the Universe. Indeed, we fall short by about a factor of 10: the average density of baryonic matter is lower than 10 percent of the critical density ($\Omega = 1$). If it were even as high as 20 percent, the amount of deuterium, helium, and lithium in the Universe would be quite different from what is observed. As we mentioned in Chapter 3, the matter lurking around galaxies also happens to be about 10 percent of the critical density, so the dark halos might well be just normal, baryonic matter that somehow fails to shine.

Something besides baryonic matter was probably needed, however, to make galaxies in the first place. As argued in Chapter 6, baryons had to be distributed far too smoothly when the Universe was about 100,000 years old for them to have coalesced into galaxies less than 1 billion years later. Something else completely unaffected by radiation had to be present during that early epoch. It was either some form of cosmic seed or some kind of dark matter that could have started clumping well *before* the baryons and radiation decoupled. Then the smooth baryonic matter would have subsequently condensed upon these preexisting seeds or clumps to form the luminous blobs and spirals we see everywhere today.

One category of cosmic seed that is unaffected by radiation includes the cosmic strings, which we introduced in Chapter 5. Such exotic, ultramassive loops and strands of primordial energy density would be sturdy relics of the GUTs epoch that somehow managed to survive for a million years or more. Another alternative would be to generate new seeds during phase transitions that might have occurred *after* radiation decoupled from matter. In any case, it is becoming patently obvious that we need something else besides normal protons and neutrons to help stimulate galaxy formation. They just cannot do it of their own accord in less than a billion years.

Finally, the introduction of grand unified theories of interparticle forces has precipitated a small revolution in the way cosmologists think about the very early Universe. Not only do GUTs provide a straightforward mechanism for generating an excess of matter (over antimatter), but they also offer the tantalizing prospect of inflation, which solves several other pressing problems all at once. But with inflation the density of the Universe must be *exactly* the critical density. Space must be absolutely flat. Given that visible matter yields less than 1 percent of the critical density, and that all baryonic matter (visible or invisible) must be less than 10 percent, inflation therefore requires that 99 percent of all matter is dark—and that 90 percent of it cannot be ordinary matter at all. The protons, neutrons, and electrons that make up our bodies and everything about us must be fairly rare and exotic particles in the overall scheme of the Universe, if inflation or something like it is true.

Having reviewed all the major cosmological and physical arguments for the existence of dark matter, we can turn now to the perplexing question of its nature. What could this dark matter possibly be? Before examining the more exotic propositions being put forth recently, we will consider baryonic dark matter one last time. After all, it probably makes up about 5 percent of the entire Universe—hardly an insignificant fraction. Nonbaryonic dark matter then falls into two broad categories, known as "hot" and "cold" dark matter, depending on whether or not it was traveling at relativistic speeds (close to the speed of light, that is) at the time galaxies began forming.

In Chapter 10 we will use these dark-matter candidates together with two species of primordial ripples that might have arisen during some cosmic phase transition—quantum fluctuations or defects in space itself (cosmic strings, for example)—to develop scenarios for the formation of galaxies and large-scale structure. To obtain structure in the Universe, we need both matter (visible and dark) *and* something to trigger gravitational collapse, just as we need a cloud of water vapor and some kind of seed to make rain. When the different kinds of seeds are combined with the various options for dark matter, they lead to distinct patterns for visible structure in the Universe. By comparing the structures observed with those that scientists can predict, it may be possible to determine which combination is the correct one.

———————————— • ● • ————————————

By their very nature, baryons (and the electrons that accompany them in atoms) should be fairly easy to spot. Being rather gregarious corpuscles, they like to interact with photons—particles of electromagnetic radiation—and do so almost every chance they get. Because of this propensity, we might naively expect all baryonic matter to be visible. What you see is what you get.

But the dark baryons in the Universe curiously outnumber those that shine. Remember that luminous matter contributes less than 1 percent of the critical density, but our Big-Bang nucleosynthesis arguments indicate that baryons make up *at least* 2 percent and possibly as much as 10 percent. Thus, the majority of baryons must occur not in shining objects like stars, but hidden away in something else like clouds of gas or dust, large planets, brown dwarfs, neutron stars, or black holes. Perhaps we can establish which, if any, of these possibilities might be true.

Indeed, recent studies of x-ray emissions by the Einstein satellite have revealed that there is a lot of extremely hot gas in rich clusters. There seems to be as much baryonic matter in this gas as in the shining stars of these ensembles. (There could be additional gas in these intergalactic spaces that is not hot enough to emit x-rays.) Perhaps the dark, gaseous matter in the halos of individual galaxies was stripped from them as they fell inexorably toward the center of the cluster. The energy of impact and the friction between galaxies would have heated the gas to temperatures of millions of degrees, so that it emits the high-energy x-rays witnessed.

But a cold, thin gas distributed uniformly throughout the halo of the Milky Way would already have been detected indirectly

—because it would absorb light coming in from other galaxies. Astronomers studying this light should have noticed some telltale dark lines in the electromagnetic spectra from these galaxies, at a few specific wavelengths characteristic of the guilty atoms. As these are not seen to any significant degree, we can say that the bulk of the stuff in our own dark halo is not a cold, uniform gas of baryonic matter.

Clouds or clumps of gas, however, would be another story altogether. They could easily exist in the halo and simply not have intercepted our line of sight to any distant galaxy whose spectral composition has been studied thus far. They might also inhabit the tremendous voids between galaxies. Indeed, recent research by astronomers seems to reveal huge intergalactic clouds of hydrogen gas. Studies of the light coming from distant quasars show a "forest" of dark absorption lines that can be explained if this light had passed through clouds of neutral hydrogen gas on its way to Earth. These clouds are comparable in size to a normal galaxy; the total amount of matter in them is estimated to be about as much as occurs in shining objects. One such cloud was recently discovered relatively close to the Milky Way, seemingly on the verge of collapse.

Some astrophysicists have advocated clouds of dust as the baryonic dark matter. But it is difficult to understand how this dust could be like the normal interstellar dust composed of heavy elements (all elements heavier than helium), the kind generated by nucleosynthesis in stellar cores and then spewed into space during supernovae. The average abundance of these heavy elements in the visible Universe is very small—only about 2 percent. It would be odd indeed if such a heavy-element dust contributed any greater percentage of the invisible halo, especially when there are very few visible stars there at all. (One loophole, however, would be hydrogen gas frozen into grains, or "snowballs"; such snowballs are hard to form and fragile, but they might well exist.) Whatever the case, absorption of incoming light by *any* kind of dust should obscure other galaxies and redden their light—just as the dust in the disk of the Milky Way does to the more distant stars within it. Such an effect is not seen to any great degree. We conclude that dust is not a good place to hide large amounts of baryonic dark matter.

Another possibility is that the baryonic dark matter may be bound up in low-mass objects called brown dwarfs, which are intermediate in mass between a planet and a star. Containing less than 10 percent of the Sun's mass, such brown dwarfs are like smoldering embers. They emit little visible light and are almost impossible to observe unless located very nearby. Even then, one has to look in the infrared portions of the electromagnetic spectrum, where they emit most of their radiation. Concrete evidence for a brown dwarf orbiting a nearby star may have been obtained recently by Ben Zuckerman of UCLA and Eric

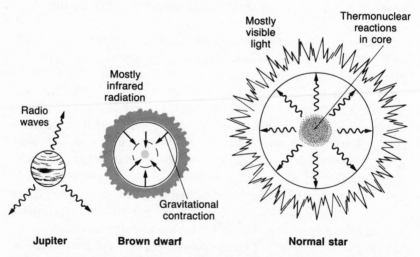

A Jupiter-sized planet compared with a brown dwarf and a normal star. A brown dwarf generates heat by gravitational contraction and emits some radiation, mostly in the infrared portion of the spectrum. Normal stars like the Sun undergo thermonuclear reactions in their cores and emit visible light.

Becklin at the University of Hawaii using an infrared telescope on Mauna Kea. In 1989, William Forrest of the University of Rochester reported strong evidence for four such objects located in the constellation Taurus.

In the same category are large planet-like objects the size of Jupiter or smaller. These are large balls of gas that, like Jupiter, are not sufficiently massive to initiate thermonuclear reactions. Any object with less than 8 percent of the Sun's mass (Jupiter has 0.1 percent) will not undergo nuclear burning, but it can generate heat and give off radiation due to gravitational contraction. As its matter sinks inward under the force of gravity, it picks up kinetic energy like a ball rolling down a hill. This energy is eventually converted into heat or radiation.

To account for all the baryonic dark matter using only brown dwarfs and Jupiters requires some unusual assumptions, however, as Dennis Hegyi at the University of Michigan and Keith Olive of the University of Minnesota have argued. There would have to be tremendous numbers of such dark, low-mass orbs in the halo but extremely few heavy objects that *could* initiate burning and hence be visible. Although different from the distribution of stars in the disk of the Milky Way, such a mass distribution — skewed so much to the low-mass end — seems very unusual at first glance, but it may not be completely impossible.

To put all the baryonic dark matter into high-mass objects like neutron stars and black holes seems equally difficult. These dark remnants of supernovae, the explosive deaths of very massive stars, could certainly hide a lot of baryons without giving up much light. But to explain *all* the baryonic dark matter this way would require that the mass distribution of any primordial objects be skewed to the high-mass end. Then, too, where are the other remnants of such paroxysms, including the substantial quantities of heavy elements that are usually disgorged? In the great 1987 supernova SN1987A, newly minted elements such as iron, nickel, and cobalt were observed in the matter thrown out into space. Unless the majority of heavy elements are somehow trapped in the black holes themselves or clumped into objects such as asteroids or planets, we would expect to notice large quantities of heavy elements in dark halos. But we see nothing out of the ordinary.

A word of caution is necessary, however, before we dismiss these possibilities. We do not really know a priori what the primordial mass distribution was for stellar-sized objects. In the early Universe there were no heavy elements around at all, so the original distribution of stellar masses was probably quite different from what appears to be the norm today. The primordial distribution could have been strongly peaked toward the low-mass or high-mass ends. Currently there is much debate on this point. Had it been peaked toward the low-mass end, then brown dwarfs and Jupiters should still be around in great abundance. If peaked at the high-mass end instead, there would be an abundance of black holes.

Furthermore, any orbs (including Jupiters, brown dwarfs, and black holes) that formed before the galactic disk would end up naturally in a spherical distribution or halo. The very old stars in globular clusters, for example, are spherically distributed about the center of a spiral galaxy, and are not found in its disk. Recent observations seem to indicate, too, that the disks themselves did not form until several billion years after the galaxies collapsed. Thus, it is not too difficult to imagine that any objects (or their remnants) located in the halo regions of a spiral galaxy are not the same kinds of things as those in the disk.

The exact location of baryonic dark matter, therefore, is still an open question. It could reside in galactic halos and even contribute essentially all the invisible mass thought to exist there. Or a substantial fraction might be coasting about in the vast intergalactic voids — either in clumps or as hot, ionized gas that is too cool to emit x-rays. If the dark baryons are located there, however, some other kind of dark matter is needed to explain the halos. Wherever baryonic dark matter is hiding, we should in principle be able to observe it (unless it's trapped in black holes) eventually, because it's the normal, familiar stuff that can emit and absorb photons. Finding it will give us solid clues about the process of galaxy formation — and may also tell us more about the nonbaryonic dark matter.

———————————— • ● • ————————————

Cosmologists distinguish two types of nonbaryonic dark matter: hot and cold. (These names were coined by Richard Bond of the University of Toronto and Alex Szalay of Eötvös University in Budapest.) Hot dark matter is composed of particles that were moving very fast — at practically the speed of light — just prior to the time of galaxy formation. Cold dark matter, on the other hand, refers to any corpuscles that would have been moving very slowly during that epoch. At one time, physicists also spoke of "warm" dark matter, which included any particles with intermediate speeds, but this has not proved to be a very useful category and has dropped from common usage.

A light neutrino has been the favorite candidate for hot dark matter. If one of the three kinds of neutrinos had a small mass of about 30 electron volts (30 eV) or so, less than one ten-thousandth the mass of an electron, then these neutral particles would have been racing about at almost the speed of light before the epoch of galaxy formation. As the Universe continued to cool, such light neutrinos would slow down until their mass-energy was greater than their energy of motion, and it would make an important contribution to the overall cosmological mass density. With so many neutrinos around — about 500 per cubic centimeter or 7500 per cubic inch — they could easily contribute enough mass to close the Universe, even if they weighed in at only 20 to 30 eV apiece.

The interest in neutrinos swelled during 1980 when two independent measurements suggested that electron neutrinos might indeed have a mass in the appropriate range. A team of physicists led by Frederick Reines at the University of California at Irvine reported observing a phenomenon called "neutrino oscillations," which can occur only if these neutrinos have a mass. Another group in the Soviet Union led by Vladimir Lyubimov claimed it had measured the mass of the electron neutrino to be about 30 eV. But subsequent experiments have repeatedly failed to confirm these results. The consensus today, based on an experiment done at the Los Alamos National Laboratory, is that the electron neutrino has a mass less than 10 eV.

This is somewhat better than the limit derived from the supernova SN1987A in the Large Magellanic Cloud. The more massive the neutrino, the more spread out in time is the intense burst of neutrinos emanating from the collapsing core of the original star. From the observed time interval of about 10 seconds, as witnessed in enormous underground detectors located in Ohio and Japan (originally set up to study proton decay), we can conclude that the electron neutrino's mass is less than 25 eV.

Even though the electron neutrino is not massive enough to close the Universe, one of its cousins from the other two quark–lepton families might well be. Perhaps the muon neutrino or tau neutrino, mentioned in Chapter 2, has the right mass to do the job. We expect that the neutrino associated with the heaviest lepton, the tau neutrino, would be the most massive—and therefore contribute most to the cosmological mass density. If so, the tight limits on the electron neutrino mass would pose no barrier.

All hot dark-matter candidates run into a severe problem, however; they have difficulty accounting for galaxy formation in any universe where quantum fluctuations in the primordial density serve as the only seeds for gravitational collapse. With speedy particles, any ripples in the otherwise smooth sea of original matter rapidly become smeared out over regions whose size equals the distance the particles could have traversed since they decoupled. If their velocities are close to the speed of light, this is the distance to the horizon at any particular epoch. In a universe made primarily of hot dark matter, therefore, the smallest region that can begin coalescing (without the help of some

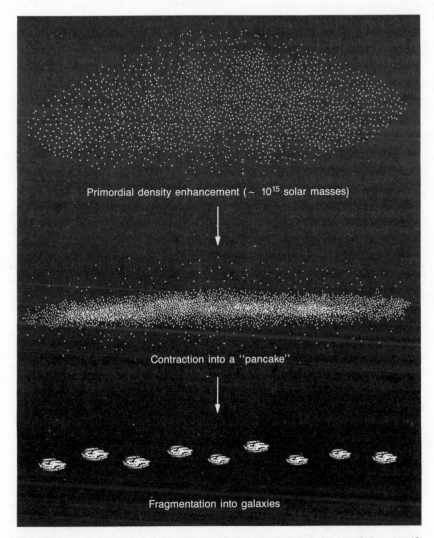

Top-down structure formation with hot dark matter. A large mass of about 10^{15} solar masses collapses first into a thin "pancake," which breaks up into clusters that eventually condense into individual galaxies.

different kind of seed) corresponds to the size of the event horizon at the time of galaxy formation.

As such a process could not even begin until the Universe became dominated by matter at an age of about 10,000 years, we obtain extremely large celestial objects, at least initially. At that time a horizon contained the amount of mass found in a rich cluster today—the mass of a thousand galaxies, or a hundred trillion suns. Therefore, we get a "top-down" scenario for galaxy formation: rich clusters condense first and galaxies form later, as these clusters break up into smaller droplets.

Such a top-down scenario strains the imagination by pushing the era of galaxy formation too late in the game. From the observation of quasars and galaxies at very large redshifts, we know that these aggregates had already begun to form by the time the Universe was only half a billion years old. To make such massive objects that early is difficult enough when it happens directly. To have them arise by a secondary process, *after* the fragmentation of a much larger entity, seems literally impossible. So hot dark matter has run into severe difficulties explaining galaxy formation—at least if the seeds inducing gravitational collapse are primordial ripples in the density of this matter itself.

If these ripples occur not in the hot dark matter, however, but in something else, there might be a way to avoid this problem. Suppose there were a mechanism that allowed galaxy-sized primordial ripples to remain confined and not be smeared out immediately by fast-moving particles. Then it would still be possible in a Universe dominated by hot matter to build up galaxies by the time it was one-third of a billion years old. Cosmic strings (or other defects in space) provide just such a mechanism, and we will say a lot more about them in the next chapter. But lacking such an additional entity, hot dark matter is not a viable alternative, because it encounters far too much difficulty making galaxies early enough in the Universe.

———————————————— • ● • ————————————————

Cold dark matter, on the other hand, is ideal for creating galaxies. These particles move too slowly for any primordial ripples in the otherwise uniform density to become smeared out; they can therefore begin clumping together of their own accord well before baryonic matter decouples from radiation — forming invisible seeds for later galaxy formation. Clumping first on small scales, particles of cold matter would rapidly build up galaxy-sized objects by the time the Universe is a few billion years old. A few objects might form earlier, perhaps as early as 500 million years.

Cold dark matter was so successful in solving the knotty problem of galaxy formation that by 1983 it had easily become the leading candidate. Detailed models of the formation process, pioneered by Joel Primack and George Blumenthal of the University of California at Santa Cruz, agreed beautifully with the observed motions and structures of galaxies. Starting out with a rotating cloud of baryons and cold dark matter mixed together indiscriminately, the baryons would interact with one another and give off light, thereby losing energy and speed. The baryons would then collapse inward toward the center of the cloud, collecting there in a flattened disk due to their rotary motion. Meanwhile, the cold dark matter stayed aloft, forming a dark halo swirling about the luminous central disk. This picture corresponds extremely well with the actual structures we observe today: flat, shiny spirals basking placidly in the midst of huge dark halos.

Fueling all the optimism over cold dark matter, too, was its early success in reproducing the large-scale structure of the Universe — at least as it was understood before 1985. Computer simulations of this structure, performed by Marc Davis and collaborators at Berkeley, worked far better if cold rather than hot dark matter was used. But more recent observations of large-scale structure and high peculiar velocities (see Chapter 7) have been posing severe problems for the cold dark-matter scenario. Although such features as the Great Wall can occasionally appear in simulations based on cold dark matter, they are not

Homogeneous concentration of baryons and dark matter (~ 10^{12} solar masses)

Baryons radiate and fall toward center of cloud

Glowing disk of stars forms at center of cloud

commonplace — as seems to be the case from the recent pencil beam surveys.

One possible example of cold dark matter would be a heavy neutrino, with a mass at least *twice* that of a proton. This particle would not be the companion of any known lepton — the electron, muon, or tau. Nor did any such particle turn up among the decay products of the Z particle in experiments at Stanford's SLC or the LEP collider at CERN. Thus, the mass of any such heavy neutrino must be greater than half the Z mass, or more than 45 GeV. Such a massive beast would not be a normal neutrino at all, if it exists; it would require a fundamental revision of the Standard Model of particle physics.

Any other heavy, neutral, weakly interacting particle would also suffice as cold dark matter. Such creatures arise naturally in grand unified theories with supersymmetry (SUSY). You may recall from Chapter 5 that these theories effectively *double* the number of fundamental particles by adding a "supersymmetric partner," or "sparticle," for every normal particle that occurs in the Standard Model. Photons have photino partners, electrons have selectrons, quarks have squarks, gluons have gluinos, and so on. The masses of these sparticles are not predicted by the theories, but they are generally believed to be greater than a proton mass and perhaps far more than that of the Z particle. SUSY particles have not yet been discovered at particle accelerators, but the enthusiasm for their existence has remained high, especially among theoretical physicists.

Before any such particle can contribute to the mass of the Universe, however, it would have to be absolutely stable, or at least extremely long-lived. It could not decay rapidly into lighter objects or pure energy — as so many subatomic particles are wont to do. Fortunately, out of the virtual armada of sparticles predicted by any particular SUSY theory, there should be

---

Left: Spiral galaxy formation with cold dark matter. Baryonic matter gives off radiation and loses energy, thereby falling more rapidly to the center of the cloud, while the dark matter remains aloft.

ELEMENTARY PARTICLES IN THE STANDARD MODEL AND THEIR
COUNTERPARTS AS PREDICTED BY SUPERSYMMETRIC (SUSY) THEORIES

Standard-Model particle	Corresponding SUSY particle
Leptons	Sleptons
neutrino	sneutrino
electron	selectron
muon	smuon
tau	stau
Quarks	Squarks
Photon	Photino
Gluon	Gluino
Weakons	Weakinos
W	wino
Z	zino
(Higgs)	(Higgsino)

one sparticle called the "lightest supersymmetric particle" (or
LSP, for short) that is indeed stable. The other, heavier sparticles
would decay into the LSP (plus normal particles), but the LSP,
being the lightest, can decay no further. The buck stops there.

The photino has most often been cited in the scientific litera-
ture as the probable LSP, with the gravitino (the SUSY partner of
the graviton, the particle that carries the force of gravity) and
higgsino (the SUSY partner of the Higgs particle) also advocated. The generic name given to this class of neutral SUSY
particles is "neutralino." Heavier than a proton, such particles
would move extremely ponderously, as required. Perhaps the
dark halo of our own Milky Way galaxy is a tremendous swarm of
neutralinos through which the Earth and Sun and all the stars
pass.

At first, the only constraint on the mass of the LSP (coming
mainly from cosmological considerations) was that it be heavier

than about 2 GeV, twice the mass of a proton. But the fact that no SUSY particles have yet been observed at particle accelerators and colliders—LEP, SLC, and the Tevatron—is pushing this limit steadily upward. If it exists, the mass of the LSP is now thought to be at least 15 GeV and is probably much more.

Another favorite candidate for cold dark matter is a wispy little particle called the "axion." When first proposed in 1978 by Steven Weinberg of the University of Texas and independently by Frank Wilczek of Princeton, it was predicted to have a mass about a fifth that of an electron and to decay in less than 1 second into two photons. Hardly a good candidate for stable dark matter. But when high-energy physics experiments soon showed such a creature could not possibly exist, other scientists revised the original theory, predicting an axion mass of less than 0.0001 eV—or less than a *billionth* the mass of an electron.

With a lifetime that is effectively infinite, such an "invisible axion" makes an excellent dark-matter candidate. It would be generated copiously in the early Universe, at about the moment quarks are freezing into protons and neutrons. Even though they would be incredibly light, these hypothetical entities would be produced so copiously—in far greater numbers than neutrinos, in fact—that they could easily dominate the total mass of the Universe. And axions would be produced essentially at rest, too, completely out of contact with all the other fast, energetic corpuscles dashing about the early Universe. Thus, they would form *cold* dark matter that could begin clumping together on small scales, well before everything else.

Another object that might have formed when quarks coalesced into baryons is called a "quark nugget." This is a hypothetical aggregate of *many* quarks, far more than the several hundred or so trapped in heavy nuclei like gold, lead, or uranium. Unlike normal atomic nuclei, which are composed only of up and down quarks, these nuggets might contain a substantial number of strange quarks, too. Arnold Bodmer at the University of Illinois, Edward Witten of Princeton, and Robert Jaffe of MIT have suggested that such a combination might actually be a stable form of matter.

Typical sizes of these quark nuggets could range from 1 milli-meter to 1 meter—all of it quarks and therefore extremely heavy stuff. If any such stable clumps could exist and be pro-duced in the early Universe, they would make an excellent candidate for cold dark matter, for they would be very heavy and slow-moving. Quark nuggets might conceivably be produced at particle accelerators, in experiments where two heavy nuclei are smashed together at very high energy.

In the same vein, we can imagine creating little black holes during one of the phase transitions of the early Universe, before the onset of Big-Bang nucleosynthesis at a time of about 1 sec-ond. Such "primordial" black holes would be smaller than the "normal" black holes that we now think form as one of the end products of stellar collapse. They might occur if there were sufficiently large density ripples at some moment in the early Universe.

To be stable, a primordial black hole would have to contain at least 1 trillion kilograms of matter—about the mass of a moun-tain. Any smaller and they would have radiated away their total mass–energy, as Stephen Hawking first showed. But a primor-dial black hole bigger than this and smaller than the mass of the Sun would make excellent cold dark matter. Perhaps at the moment when quarks froze into protons and neutrons, there were density ripples large enough so that many, many quarks became glued together into objects with the mass of a planet. Such primordial black holes might now be lurking unseen in the dark halo of our own galaxy.

In truth, quark nuggets and primordial black holes are another kind of baryonic dark matter, because they can be produced from quarks that do not shine. To make them, we do not have to propose anything more exotic than the quarks themselves, which must have existed in profusion during the early Universe. Instead, we only need to establish unusual "states" of quark matter that might have been generated *before* Big-Bang nucleo-synthesis occurred, thereby taking a sizable fraction of the quarks or baryons out of circulation at a key moment in cosmic evolution. Nucleosynthesis can then proceed, converting the remaining protons and neutrons into hydrogen, helium, and lithium.

Topological remnants from the GUTS epoch are another possible form of cold dark matter. These would be exotic objects like the cosmic strings or magnetic monopoles mentioned in Chapters 5 and 6, which did not undergo a transition from the unified, symmetric space that existed before 10^{-34} second to the ordinary space in which we live today. As the energy density was tremendous in that epoch, these would be extremely massive objects—far, far more massive than normal particles. Small loops of string, if they were stable (and this is a *big* if), would act like cold dark matter. So would magnetic monopoles, which are like tiny balls of symmetric space instead of tubes or loops. Because of their tremendous masses, they too would move extremely slowly. Although inflation supposedly rids the Universe of almost all magnetic monopoles, it is conceivable that a new version of the theory might leave some such remnants that could survive to the present day.

Although cold dark matter is very proficient at galaxy formation, however, it falls short in other important areas. Because it is moving so slowly to begin with and clumps together so quickly, it does not attain very high velocities. Thus, the hypothesis of cold dark matter has a difficult time explaining the high peculiar velocities that have been witnessed recently in our own corner of the Universe.

Yet another problem is that cold dark matter would clump together with baryons to make visible galaxies and halos. Although this reproduces the observed structure well, it causes severe problems for inflation, which requires a value of Ω *exactly* equal to 1. If you remember, the internal dynamics of galaxies and clusters can support values of 0.1 to perhaps 0.3, but not 1. If cold dark matter and baryons were clumped together in the same proportions everywhere, that would mean that Ω is no larger than 0.3. We therefore need something *else* besides cold dark matter if inflation is to hold true.

Advocates of cold dark matter usually insist on what is called "biasing" to resolve this problem. In this approach, for some ad hoc reason, most of the clumps of baryons and cold dark matter simply do not shine at all. Only the fortunate 10 percent or so, perhaps the most massive clumps, can ignite the baryonic matter at their cores. The other 90 percent remain dark and invisible, perhaps populating the huge voids in intergalactic space as "stillborn galaxies." We can compare this situation to a mountain range in which only the highest peaks are snow-capped in summer. From far away, these would be the only ones visible.

———————————— • ● • ————————————

A final dark horse candidate is what has become known as "shadow matter." In any Theory of Everything, which purports to unify gravity with the other three fundamental forces at times before 10^{-43} second, there can arise a parallel universe (commingled with our own) that has completely different forces and particles. This is not one of the "bubble universes" that arise in new inflation shortly after 10^{-34} second; those are distinct spaces disconnected from ours. It is a shadow universe that occupies the very same physical space as the familiar Universe but has no normal interaction with it other than through the force of gravity. We can imagine that the particles of shadow matter might form shadow atoms and molecules. There could be shadow rocks and plants, even shadow people, planets, stars, and galaxies that would pass right through our own almost completely unnoticed.

Once the two become divorced from one another after 10^{-43} second, however, the only influence this shadow universe would have upon our own would be through the force of gravity. All matter in the Universe attracts all other matter, no matter what kind. Depending upon whatever masses and speeds any such shadow matter assumes, it could act like either hot or cold dark matter as seen from our own perspective. If the shadow particles were heavy or slow, they would behave like cold dark matter and could clump with baryons on galactic scales. If they were light and fast, they would behave like hot dark matter and aggregate only on the scales of clusters or larger. From a cosmological viewpoint, it does not matter how a particle arises. If it moved slowly at the time of galaxy formation, it falls into the category of cold dark matter; if it moved quickly, it is hot dark matter.

Shadow matter, if it exists, cannot be exactly like the stuff in our own, familiar Universe — made of shadow quarks and shadow leptons that come in three shadow families, each with its own light shadow neutrino. That arrangement would violate the nucleosynthesis limit on the number of light subatomic particles (see Chapter 4). Remember, we already have three different types of light neutrinos, and Big-Bang nucleosynthesis arguments have difficulty accommodating even one more kind of light particle. Other than this restriction, however, we have no knowledge at all about the properties of shadow matter. And because it could only interact with us through gravity, we would have a hard time measuring them!

———————————— • ● • ————————————

In this chapter, we have tried to describe the most popular candidates for dark matter. This is by no means an exhaustive list, however. Virtually every theorist has his or her own pet theory, and the dark particles it predicts may not be included above — or they may be a variation upon one of the general categories presented. Because very little is known about the properties of dark matter, other than the fact that it exists, there is plenty of room for theoretical speculation right now. As measurements improve, this situation should change, and some or all of these candidates will inevitably fall out of favor. Quite possibly, dark matter will turn out to be something completely different from anything mentioned above.

10

Seeds of Collapse

Another fossil relic remaining from the Big Bang is the large-scale structure of the Universe itself. Its importance to a complete theory of genesis has begun to be understood only in recent years, as one surprising observation after another has come to light. The foamy, frothy structure with its voids and sheets, the strong spatial correlations between clusters, and the high peculiar velocities of galaxies in our surrounding cosmological neighborhood have sorely challenged previous ideas about the early moments of existence. And we can be sure that the surprises have not ended yet.

Simply specifying the kinds of matter in the Universe — baryonic or not, cold or hot — is not enough to explain the striking patterns observed in the heavens. We need more information than this. There is clearly no shortage of dark-matter candidates running for election, as we have been discussing throughout this book. But any single ingredient by itself fails to account for at least one important aspect of the large-scale structure. The ultimate solution has to be more complex.

To build up the Universe with all its apparent structure, we also need some kind of primordial seeds spanning the cosmos upon which matter can condense, like water vapor into dew-drops upon a flimsy spider web. These seeds might be little more than a random pattern of overdensities — tiny ripples in an otherwise perfectly smooth ocean of original substance. Or they could be extended physical remnants of the symmetric GUTs phase of the early Universe such as cosmic strings, or perhaps domain walls and other more complex features.

In fact, all current ideas for generating such cosmic seeds involve new physical processes or entities that cannot be studied in terrestrial laboratories. By interpreting the large-scale structures observable today, the outward manifestations of these Big-Bang remnants, we may be able to learn about the physics of unification, which must have governed the events that occurred in the early moments of the Universe. Telescope observations of the largest structures in the Universe, that is, may teach us something about fundamental physics. This is an extremely exciting prospect.

———————————— • ● • ————————————

Until recently, a random pattern of overdensities was the most popular kind of cosmic seed advocated by cosmologists. In theories with inflation, they arise naturally during the GUTs phase transition — as quantum fluctuations on a smooth primordial sea of matter and energy. In fact, such ripples randomly distributed about space are an almost unavoidable consequence in the theory of new inflation. We do not have to impose any ad hoc assumptions in order to generate them. As noted in Chapter 8, we can even rule out specific grand unified theories, such as SU5, based on the fact that overdensities that were generated in the GUTs transition would have been far too large. Instead of contracting very gradually to produce galaxies, they would have collapsed immediately into tremendous black holes, and we would not be here to tell about it.

Nowadays cosmologists take these random overdensities pretty much for granted. But before the 1980s any such primordial ripples were strictly ad hoc assumptions about the initial conditions of the Universe; there was no conceivable way to generate them. The inflation hypothesis fortunately came along to supply a mechanism that could in principle produce *both* the smoothness at extremely large distance scales and the small-scale ripples needed to form structures. Although currently available theories require some fine tuning to keep the ripples from becoming tidal waves, at least we now have a likely physical mechanism to produce them in the first place.

Ripples generated during the GUTs transition would have survived, in one form or another, until the epoch of galaxy formation began thousands of years later. This process cannot start until matter begins to dominate over radiation as the major component of mass–energy in the Universe. Gravity cannot enhance the ripples if radiation (particles moving at the speed of light) keeps smoothing things out. Nonbaryonic dark matter can begin aggregating once its energy density exceeds that of the surrounding radiation; this becomes true when the Universe is about 10,000 years old. But the baryonic matter cannot follow suit until it completely decouples from the radiation. For a universe with the critical density ($\Omega = 1$), this moment occurs when it is about 100,000 years old.

Until this decoupling occurs, the primordial ripples would have served only as nucleation sites where nonbaryonic dark matter could have begun to collect. This process would have continued along unobtrusively for about 100,000 years, until baryons decoupled from the radiation and themselves began to condense. Finally, after several hundred million years, luminous objects should have emerged and begun shining when these large clumps of matter started to undergo gravitational collapse. In such a sequence of events, the final pattern of observable light should in some way reflect the original random pattern of ripples that emerged from the inflationary epoch — unless other physical processes interfere.

It takes such a long time to form visible objects because the tiny ripples in the original density of the Universe can grow only slowly with the Hubble expansion. From the observed uniformity of the cosmic background radiation, we realize that these ripples had to be very small. Only when a ripple has grown to the point where the density inside it is about twice the surrounding average can rapid gravitational collapse truly begin. The entire process requires many millions of years, and sometimes billions.

As mentioned in Chapter 9, hot dark matter would have smeared out any primordial overdensities to extremely large distances. In this scenario, extremely large celestial objects possessing the masses of superclusters today would have collapsed first, to be followed by their fragmentation into smaller objects such as ordinary galaxies. Although cosmologists can simulate the apparent large-scale structure using hot dark matter condensing upon random overdensities, they hit a brick wall trying to generate galaxies early enough to correspond to the actual observations. In such a top-down scenario of structure formation, individual galaxies appear much later than observed — billions of years after the Big Bang.

Cold dark matter does a better job of making galaxies early in the game. But it clumps with baryonic matter, and we know from Chapter 3 that the matter clumping with visible galaxies and clusters can give us an Ω of at most 0.3. Thus, models of the Universe based on cold dark matter have trouble obtaining the $\Omega = 1$ required by inflation. Either we have to abandon inflation altogether (an alternative most theorists are loath to accept), or we must introduce an additional assumption such as biasing. Somehow only 10 to 30 percent of the original clumps have begun glowing, according to this line of reasoning, whereas the remaining 70 to 90 percent remain dark and hidden from view, perhaps populating the immense voids recently discovered in intergalactic space.

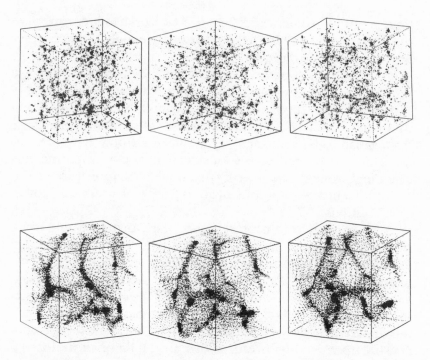

Computer simulations of structure formation in the cold dark-matter (top) and hot dark-matter (bottom) scenarios (assuming random overdensities act as the seeds). Galaxies form first and cluster later in cold dark-matter models; with hot dark matter, by contrast, clustering occurs first at large scales, followed later by fragmentation and galaxy formation.

The earliest proponents of cold dark matter with its requirement of biasing, surprisingly enough, were cosmologists with strong backgrounds in traditional optical astronomy. Groups of astronomers from the University of California at Berkeley and at Santa Cruz were particularly enthusiastic advocates of cold dark matter. They were effectively saying that they could not believe what they were seeing with their very own eyes. In other words, "telescope observations give us a distorted picture of the Universe."

Although they had resounding success in explaining galaxy formation, advocates of cold dark matter now have severe difficulty explaining how the peculiar velocities of galaxies can possibly be so high. You may remember from Chapter 7 that nearby galaxies, including our very own Milky Way, are moving at speeds close to 600 kilometers per second relative to the uniform cosmic background radiation. How can this happen?

If we postulate cold dark matter, with or without biasing, we end up with a lot of slow-moving clumps just hanging about taking up space, doing nothing very peculiar, merely expanding gradually with the uniform Hubble flow. The more biasing we put into such a model Universe, in fact, the *lower* the peculiar velocities come out. Cold dark matter seems inconsistent with high velocities.

In a general model with biasing, one can assume that luminous matter seen in telescopes and detected by radio antennas just does not reflect the underlying distribution of unseen matter. The visible matter is here, say, and the dark matter is way over there. One can certainly imagine that there is a very different structure in the dark universe than in the visible, such that we cannot base any conclusions about the underlying structure upon what we see in the heavens. Light does not *have* to trace matter.

But biasing makes it even harder to obtain high peculiar velocities. And while such velocities can occur in cold dark-matter models of the Universe without biasing, these models cannot achieve $\Omega = 1$; the biasing factor needed to fit the observations of low Ω on small scales cannot accommodate high velocities. If you begin with sluggish stuff, you end up with sluggish stuff— unless there is something *else* in the picture to impart the speed. Hot dark matter, being extremely speedy to begin with, does not have as much difficulty attaining these high velocities (but it still has the old stumbling block of making galaxies early enough, if random overdensities are the only available seeds).

Defenders of cold dark matter argue that the high velocities observed may be relatively rare in the Universe — just a freak of nature. They point out that some of their models (those without biasing) occasionally produce small regions with high peculiar velocities. Perhaps we just happen to be located in a very special part of the Universe.

Whether one employs hot or cold dark matter, however, *any* scenario based on random overdensities runs into difficulty if clusters turn out to be strongly correlated with one another — if, that is, clusters are found more often near other clusters. Unless there is something else at work, we naively expect that clusters would be randomly sprinkled throughout the vastness of space. Over the tremendous distances involved, gravity just would not have had enough time to pull them together. Even if it had a peculiar velocity of 1000 kilometers per second, a cluster could only have moved at most 30 million light-years since it formed. Although the early data on cluster correlations are preliminary, and much more study is needed, serious doubt has already been cast upon these random density ripples. They may not be the principal cause of large-scale structure in the Universe.

Another problem is caused by the observation of quasars at high redshifts, which correspond to very large distances. The greater the distance to a quasar, the earlier it was formed. Quasars have recently been observed with redshifts that indicate they were formed before the Universe was 1 billion years old. Any model that employs small density ripples as seeds has difficulty generating compact objects this early. Observations of one or two quasars with high redshifts can be dismissed as statistical fluctuations. But if they are found to be ubiquitous, then all models (whether based on hot or cold dark matter) that are seeded by small density ripples will have to be ruled out.

———————————— • ● • ————————————

The recent history of cosmology has been a fascinating period of frequently shifting loyalties, with scientists climbing onto one bandwagon after another. Until 1980, everybody seemed fairly content to concoct their universes with baryons alone, leading to an Ω of 0.1, but then along came inflation to upset the applecart with its requirement that Ω be equal to 1. At the time, grand unified theories (plus some preliminary laboratory measurements) made massive neutrinos the fashion, so hot dark matter became the favorite way to supply the extra mass needed to make up this difference. But when hot dark matter (combined with the random overdensities arising in new inflation) encountered problems with galaxy formation, physicists began deserting that sinking ship for cold dark matter.

Now that the observations of large-scale structure, high peculiar velocities, and quasars with large redshifts are casting doubt on cold dark matter, loyalties are shifting once again. The problem may not lie in the choice of dark matter, however, but in the cosmic network of gravitational seeds needed for structures to begin coalescing. During the late 1980s, cosmologists became very interested in a variety of extended features that might have been formed as the Universe went through phase transitions during its early evolution. If they survived a sufficiently long time after the phase transition, these remnants could also have served as seeds for subsequent gravitational collapse.

A common feature of phase transitions, especially those that happen rapidly, is the formation of "defects" in the material remaining afterwards. In freezing water to make an ice cube, for example, you rarely (if ever) get a single, perfect crystal of ice. Instead, the cube has a complex network of fine white lines and planes inside it. These are defects or imperfections in the crystal structure, where the individual crystals that began forming independently eventually met one another and ceased growing. Because the regular lattices of two adjoining crystals are almost never in alignment with one another, imperfections usually arise at places where the two meet.

In principle, any phase transition occurring in three dimensions can have three kinds of permanent defects, which differ according to their geometrical shape, or "topology." Defects can be point-like (zero-dimensional) monopoles, they can resemble (one-dimensional) strings or filaments, and they can assume the form of (two-dimensional) sheets or walls. Combinations of these defects and transient three-dimensional clumps known as "textures" can also occur.

Phase transitions in the early Universe, which occur when a unified, symmetrical force breaks down into two or more distinct forces, can lead to all types of topological defects in the three-dimensional fabric of space. In these unified theories, space is not an empty, formless void; it can have knots, twists, and cracks in its very makeup. These defects can arise during the phase transitions, when the different portions of space "crystallize" independently and subsequently meet. At the moment of the transition, the energy density inside the defect is about the same as that of the surrounding space, just as it is in an ice cube. As time elapses, however, the Universe expands and cools, lowering the energy density outside the defect while it remains the same inside. The interior of such a defect is a tiny remnant of the symmetric, unified space that existed when the phase transition began. It would be a piece of the original space that never managed to "cool."

The phenomenon of symmetry breaking, which is a central feature of modern unified theories of fundamental forces, leads to several different phase transitions in the early Universe, each of them capable of generating topological defects. At the very beginning of time, the hypothetical Theory of Everything breaks down into gravity and the GUTs force. This transition comes at about 10^{-43} second, when the temperature is a truly blistering 10^{33} degrees — corresponding to an average particle energy of 10^{19} GeV. In other words, every particle has the kinetic energy of a Lear jet! Next comes the GUTs phase transition (which we discussed in detail in Chapter 8) at about 10^{-34} second, and following that is the electroweak transition about 10^{-12} second later. By this time the temperature has fallen to a mere 10^{15} degrees, or a few hundred billion electron volts per particle. Finally, at about 1 microsecond after creation, the hot plasma of quarks and gluons freezes into a sea of protons and neutrons that shortly thereafter enters the era of nucleosynthesis. And there is growing interest in a possible "late-time" phase transition that

PHYSICAL CHARACTERISTICS OF THE PHASE TRANSITIONS THOUGHT TO
HAVE OCCURRED IN THE EARLY UNIVERSE

Phase transition	Time of occurrence	Temperature (°K)	Mass inside event horizon
Theory of Everything	10^{-43} second	10^{33}	10^{-5} gram
GUTs	10^{-34} second	10^{27}	1 kilogram
Electroweak	10^{-12} second	10^{15}	10^{25} kilograms (mass of earth)
Quark–baryon	10^{-6} second	10^{12}	10 solar masses
Late-time	$>10^5$ years	$<10^3$	$>10^{17}$ solar masses

might have occurred after the moment of decoupling, when the
temperature of the neutrinos falls below 1000°K.

At a particular phase transition, the size of the physical defects
that can be formed is limited by the size of the event horizon at
that moment. Events occurring at the end of a transition, like
inflation, can enhance this size substantially. Or they can make a
defect essentially nonexistent — as inflation does to the population of magnetic monopoles. The total amount of mass within an
event horizon ranges from about 10^{-5} gram when the Theory of
Everything breaks down to about a few solar masses at the
quark–baryon phase transition and to a mass roughly equal to
that of the Great Wall at the moment of a late-time transition.

———————————————— • ● • ————————————————

Cosmic string is one kind of topological defect that got a lot of
attention during the late 1980s. It is an "exotic, invisible entity
spun by theories of particle physics," writes Alex Vilenkin, a

Soviet physicist now working at Tufts University and a leading proponent of the idea. "Strings are threads remaining from the fabric of the newborn Universe." They provide an attractive network of seeds for the condensation of galaxies and clusters.

If they exist, these strands of primordial substance would be traces of the ultrahigh-energy, symmetric vacuum of the GUTs epoch that did *not* become transformed into ordinary "empty" space. In effect, they would be cracks in space. Neil Turok, an astrophysicist at Princeton who did much of the original work simulating the behavior of cosmic strings, likens them to features that arise when molten metal cools down too rapidly. "As you cool it, you find defects and cracks," he observes. "As the Universe cools, one would find that it becomes filled with this array of defects." Cosmic strings are filamentary fractures in space filled with the ultrahigh-energy residue of the GUTs epoch.

If they do exist, cosmic strings would be massive in the *extreme*. Estimates vary substantially, but they should typically weigh in at more than 1 million billion tons per centimeter, even though they are less than one-trillionth of a trillionth of a centimeter across. Pretty amazing stuff. With such a huge mass they could not help but attract nearby matter vigorously. So a primordial network of cosmic strings might well supply the necessary seeds for structure formation in the early Universe.

Unlike ordinary, everyday thread, a cosmic string cannot end (except on a magnetic monopole, which we think are extremely rare). So it either closes back on itself to form a loop, or it stretches all the way to infinity in both directions — as an "infinite" string. Although extremely massive, strings would be stretched very taut and would zoom by at nearly the speed of light, bending and buckling as they whip about. If two strings cross one another, however, or if a single strand crosses back upon itself, they can snap and then reconnect in a different manner. "Long, coiled strings cross themselves many times over," according to Vilenkin, "and closed loops get lopped off at the intersections."

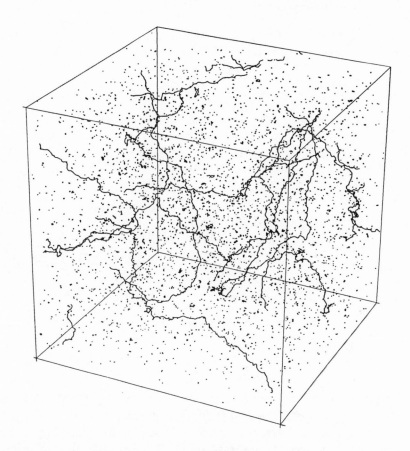

Computer simulation of a possible network of cosmic strings. Such extremely massive, one-dimensional defects in space itself may have been the "seeds" needed to trigger galaxy formation.

These loops were originally proposed as the cosmic seeds that led to the formation of galaxies and clusters. "The smallest loops will give rise to galaxies — they just sit there and attract the matter toward them," noted Turok. "Bigger loops will just collapse smaller loops onto them and form clusters of galaxies." If there is dark matter lurking around, such processes would

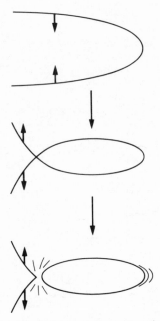

A loop of cosmic string, typically hundreds of thousands of light-years across, can be formed when a long string crosses back upon itself.

begin after matter became more common than radiation, at about 10,000 years. Structure formation would be well along by the time visible, baryonic matter decouples from radiation and joins the collapse thousands of years later. Smaller loops of cosmic string evaporate away by emitting gravitational radiation. The smallest loops evaporate quickly and the larger ones later on, with the very largest loops still in existence today.

But the failure to observe the effects of this gravitational radiation (which we will explain in more detail in Chapter 11) has already become a problem for the earlier models of cosmic strings. Some cosmologists have circumvented this problem by suggesting that the "wakes" formed by cosmic strings as they zoom through the distribution of matter are what serve as seeds. Such secondary ripples in the matter density, say Jane Charlton at the University of Arizona and David Bennett of Princeton, could induce the formation of galaxies and clusters. More recent simulations support this picture.

Cosmic strings are fairly sturdy objects, as far as things go in the blistering heat of the early Universe. Because they have a life of their own, and are not just overdensities in a primordial sea of matter and energy, they cannot be smeared away by the action of fast-moving subatomic particles. Thus, they can work well with hot dark matter as well as cold, promoting rapid galaxy formation while allowing a value of Ω equal to unity. All by themselves, cosmic strings cannot give $\Omega = 1$; something else, like massive neutrinos, is needed to do this. But they can provide a way to make galaxies early enough to correspond to our observations.

The combination of hot dark matter and cosmic strings may be one of the few ways to produce the high peculiar velocities witnessed by the Seven Samurai and other groups (see Chapter 7). High velocities will arise if the strings are extremely massive, which they would necessarily have to be in order to attract the frenzied hot matter efficiently. In such an event, the visible, baryonic matter would attain very high speeds as it fell toward the string (or its wake) during the process of gravitational collapse.

If we combine strings with listless cold matter, on the other hand, we end up with velocities too low to match observations. In this case, the string cannot be too massive, or structures would form so rapidly that the microwave background radiation would be unduly distorted. But a less massive string means weaker long-range attraction, and lower peculiar velocities are therefore generated as the matter falls inward.

If the early indications of high peculiar velocities prove true, and they are found elsewhere in the Universe, then cosmic strings may have resuscitated hot dark matter as a viable candidate. Perhaps the dark matter really is massive neutrinos, after all, clumping onto very massive cosmic strings. We should reserve judgment, however, until detailed simulations are completed with models that employ wakes of cosmic strings plus hot dark matter. The constraints imposed by the absence of gravitational radiation and the lack of inhomogeneities in the background radiation provide important practical tests of such models.

————————————— • ● • —————————————

If topological defects such as cosmic strings provided the seeds for structure formation, we can more easily explain the spatial correlations observed between galaxies and clusters. From Chapter 7, you remember that galaxies are more likely to be found near other galaxies (as compared with a random distribution), and that a similar statement is true for clusters. If features with the geometry of strings or walls were responsible for seeding the cosmos, the patterns observed in the heavens should not be random at all; they should be related in some way to the distribution of primordial defects. One would naturally expect to find two galaxies (or two clusters) close together *more* often than normally encountered in a random pattern—as is observed. Neil Turok has shown, for example, that if galaxies condense upon loops of cosmic string, then the appropriate mathematical behavior of the cluster correlation function (as determined by Bahcall and Soneira) can be obtained. Similar analyses are being made for galaxies that condense in the wakes of cosmic strings.

Another way to comprehend the patterns formed when topological defects are the seeds is through the use of "fractals." As developed by Benoit Mandelbrot in *The Fractal Symmetry of Nature*, fractals are the best way to describe the fractional filling of space. Regular geometrical shapes have whole-number dimensions: a line has one dimension, a flat disk has two, a sphere has three. In principle, however, there can also be shapes with noninteger, or *fractional* dimensions, too, hence the term *fractal*. Most of the objects encountered in nature, Mandelbrot notes, are better described using fractals as a basis than by using simple geometrical forms like triangles or spheres. Computer-graphic portraits of natural scenes — such as trees, rivers, clouds, or mountains — are now generated almost exclusively using fractals.

From the observed correlations of galaxies and clusters, it can be shown that they appear to follow a fractal pattern with a dimension of about 1.2. Almost a linear pattern (which would have a dimension of 1.0), that is, but not quite. The dimension 1.2 equals the 3 dimensions of normal space minus 1.8, which is approximately the power-law behavior ($1/r^{1.8}$, where r is the separation between two galaxies or clusters; see Chapter 7) of the correlation functions. As Turok has shown, loops of cosmic string can indeed yield fractal patterns with a dimension close to this number and give the observed degree of correlation. (They are in excellent agreement with cluster–cluster correlations, and in fairly good agreement with galaxy–galaxy correlations. Gravity probably has had some influence in the latter case, however, moving the galaxies around and affecting the original correlations.)

With topological defects, therefore, we are making encouraging strides in explaining the large-scale structure of the Universe. Loops of cosmic string seem to reproduce the observed cluster correlation functions better than models seeded by random overdensities, and in certain cases can give high peculiar velocities. Combining these defects with hot dark matter also provides a natural form of biasing — in which visible objects do not trace the underlying distribution of matter. At the time of galaxy formation, most of the hot matter was still moving too fast to clump with anything, so the slow-moving baryons would have been more strongly influenced by massive defects. This gives us a relatively simple mechanism to help explain why the density of matter near galaxies and clusters, yielding an Ω of 0.1 to 0.3, is about the same as the maximum baryon density ($\Omega = 0.1$) permitted by Big-Bang nucleosynthesis. Finally, the gravitational disturbances generated by walls or in the wakes of long cosmic strings could spawn the enormous sheet-like structures like the Great Wall discovered by Geller and Huchra. Perhaps all the

voids, sheets, and filaments recently observed are just the prog-
eny of these massive cosmic defects.

———————————— • ● • ————————————

A clever idea that achieved brief prominence in the late 1980s
was the possibility of "superconducting" cosmic strings. Instead of attracting the surrounding matter, suggested Edward
Witten in 1985, this kind of cosmic string might "explode" and
blow it away. In particular grand unified theories, he noted,
cosmic strings can conduct extremely large electric currents
because they have no resistance whatsoever; they would be
superconducting. Normally the current moving in one direction
along the string would be the same as that moving in the other
direction, so the two should cancel one another out, and there
would be no net effect upon the outside world.

If there had been a primordial magnetic field in the early
Universe, however, a loop of superconducting cosmic string
whipping about through this field would develop a net current
in one direction or the other and begin to emit electromagnetic
radiation. The loop would shrink, the current inside would
grow without limit, and the loop would shine more and more
brightly. Eventually the radiation from it would become absolutely tremendous, according to Witten's proposal. Being emitted continuously, rather than in a one-shot blast, it could drive
matter over vast distances.

Independent of Witten, and well before he proposed his idea,
Leonard Cowie and Jeremiah Ostriker of Princeton had been
advocating primordial explosions as the origin of large-scale
structure. In this approach there had to be some kind of object
that exploded during the era of galaxy formation, generating
shock waves that pushed matter violently about. Driven by these
shock waves, matter would have begun clumping into galaxies
and clusters, leaving behind huge voids in space centered on the
original points of the explosions.

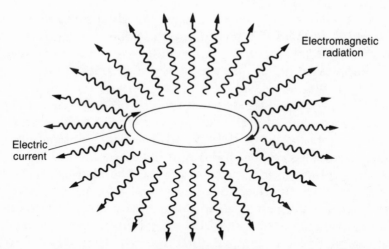

A loop of superconducting cosmic string. Because of the lack of electrical resistance, tremendous currents might build up in such a loop, generating intense radiation that would push matter outward in all directions.

Although it was an appealing explanation of the frothy large-scale structure observed, particularly the huge voids that began to be observed in the 1980s, the Cowie–Ostriker model offered no suggestion about what the exploding seeds might have been. Supernovae and other familiar astronomical explosions do not generate anywhere near enough energy to push matter through millons of light-years. Here is where Witten's superconducting strings came to the rescue.

In 1986, Ostriker joined forces with Witten and Chris Thompson of Princeton to publish a paper that incorporated both these ideas. In their model, loops of superconducting string provided the necessary seeds for the explosions, throwing off vast showers of photons in all directions. These explosions would have blown "bubbles" in the hot primordial plasma of baryonic matter and radiation. But any dark matter—hot or cold—would not have been affected at all by the explosions, for by definition it cannot interact with photons. So dark matter would have remained behind *inside* the bubbles, while the baryonic matter was pushed along through space upon their expanding surfaces like bits of flotsam before a gigantic wave. Eventually this baryonic matter would have condensed, when two bubbles collided, into the sheets of galaxies witnessed in the late 1980s.

Such a hybrid model, based upon exploding cosmic strings and a background of dark matter, incorporated many desirable features. Not only did it yield the frothy, foam-like structure of galaxies and clusters, but it did so using *either* hot or cold dark matter. Exploding cosmic strings provided another natural biasing scheme because the explosions would have driven baryonic and nonbaryonic matter far apart from each other. The model yielded a relatively smooth distribution of dark matter, whereas the baryonic matter clumped into clusters and galaxies. We still needed dark matter, of course, to obtain $\Omega = 1$. But we could generate high peculiar velocities with *either* hot or cold dark matter, because it is the explosions, not the force of gravity, that induces these high velocities. And the cluster–cluster correlation function turned out correctly, too, with a fractal dimension near 1, because the actual origin of the structure was based on a one-dimensional, string-like entity. All in all, superconducting cosmic strings were a very appealing idea.

But the extremely violent behavior expected of them has recently been used to eliminate superconducting cosmic strings as serious candidates for seeding structure formation. The explosions had to be so large and powerful that they would have perturbed the spectrum of the microwave background radiation. Instead of the almost perfect blackbody spectrum measured by the COBE satellite (see Chapter 6), that is, there would have been additional contributions. These are not observed. The only explosions still compatible with the precise COBE results would have been too weak to produce most of the striking large-scale structures that have been observed. Superconducting cosmic strings have therefore been ruled out as of early 1990.

One problem with cosmic strings and other topological defects is a potential conflict with inflation. If the inflationary episode occurs after the formation of the cosmic strings, or any other topological defect that arises during the GUTs epoch, then we would expect them to be inflated away—just like magnetic monopoles. There would simply not be enough of them left over afterward to provide the many primordial seeds needed for galaxy and cluster formation. Although we can tinker with grand unified theories so that strings form *after* inflation, or at least at the very end of this episode, such an ad hoc approach is not very convincing. There are theories called "extended inflation" in which the defects are not inflated away, but these ideas have not (as yet) gained wide acceptance.

Another way around this difficulty is to generate the necessary seeds at another phase transition that occurs very late in the Big Bang. If it happens after radiation decouples from matter, it would not disturb the uniform distribution of cosmic background radiation at all. Such a late-time phase transition might occur, for example, if neutrinos have a very tiny mass of about 0.01 eV, which is 20 billionths of the electron mass. If this mass were due to a symmetry-breaking mechanism involving a Higgs-like field, then it would lead to a late-time phase transition.

Because this transition occurs *after* matter and radiation decouple, any topological defects it generates should have a negligible effect on the distribution of microwave background radiation today, even if it induced very large ripples in the matter density. A late-time phase transition could also produce strings, textures, and domain walls (often called "cosmic membranes"), which are two-dimensional sheets of vacuum energy density. Such large density ripples or topological defects could begin forming structures almost immediately, without having to wait through a period of slow linear growth as discussed in Chapter 6.

In this scenario, the large-scale structure of the Universe is closely linked to the seeds emerging from the phase transition. Like strings, both textures and domain walls could be attractive or exploding; thus, they can produce all sorts of interesting structures and velocities. And because the structures can form very rapidly afterwards, late-time phase transitions might also be able to explain the existence of objects like quasars observed recently at extremely high redshifts. More conventional models involving random overdensities have a very difficult time producing large numbers of these objects so early in the evolution of the Universe.

The idea of late-time phase transitions is so recent, however, that cosmologists have not yet had time to explore many of its ramifications. We need to simulate the evolution of textures and cosmic membranes, for example, and see if they can indeed generate the observed features of galaxies and clusters. If further measurements continue to indicate that the microwave background radiation is extremely smooth, no bumpier than a few parts per million, then theories of this type may be the only ones that can remain viable. What is now considered a radical idea might have to be taken very seriously.

If borne out by further studies, the apparent regular succession of "great walls" discovered in the pencil-beam surveys (see Chapter 7) might well be evidence for a late-time phase transition. The typical size of these walls — several hundred million light-years — is about what we would expect. And such repetitive, "crystalline" kinds of structures can be directly associated with phase transitions (rather than the result of gravitational attraction), just as crystals grown in a laboratory are the result of a transition from the liquid to the solid phase. This exciting possibility, which cropped up in early 1990, is sure to be the subject of intense activity in the coming years.

———————————— • ● • ————————————

We came a long way during the 1980s toward understanding the origins of structure in the Universe, but we may still have a long way to go. Each of our scenarios has a number of solid successes and some unresolved difficulties. Perhaps there is yet another alternative for the primordial seeds that we have not even begun to consider. Cosmology is in such a state of flux at present that it would be wholly unwise to exclude, a priori, such an eventuality. Surely there are more new fashions and rich, bizarre theories waiting for us just down the road.

11

•

Probing the Cosmos

*T*oday we have strong evidence that the bulk of the Universe is made of dark matter unlike anything noticeable in everyday life. Physicists accept this proposition as almost an established fact. But we find it difficult to explain all the intriguing structural features that you have encountered in this book merely by invoking one of the many possible candidates. The full explanation must be more complex than that, involving not just dark matter but something *else* like ripples in the primordial ocean of matter, some kind of biasing, or topological defects in space itself. How to distinguish among all the available alternatives? How can we determine the complete answer?

Faced with similar problems, cosmologists of an earlier day would have shrugged their shoulders in surrender and retreated to the safety of their equations, convinced that measurements were next to impossible. But not any more. Recent advances in science and technology have put cosmology on an equal footing with other sciences. It has now become an *experimental* science with real-world consequences that can help us test its proliferating hypotheses.

While traditional astronomers are busy scanning the electromagnetic spectrum for odd features required by the various theories, a new breed of high-energy astrophysicist is studying ultrahigh-energy cosmic rays and neutrinos coming from space. All these scientists have many new tools at their disposal, both in space and on the Earth, to help them seek answers to cosmological questions. Particle physicists have taken up the hunt for dark matter, too, seeking its telltale footprints in the surrounding halo of the Milky Way or trying to produce it by brute force in experiments performed at powerful colliders around the globe. There is plenty of activity going on now, or about to commence, that should provide much-needed data in the near future.

—————————————— • ● • ——————————————

An important question remaining is the location of the dark baryons, which must be hiding *somewhere* out in space. As we know from Big-Bang nucleosynthesis arguments, the total density of baryons is substantially greater than the portion we can presently see. Are the dark baryons haunting the galactic halos? Or do they lurk deep within the intergalactic voids? Even though they are supposedly dark, these dim baryons may still have some kind of telltale signature that would enable astronomers to locate them.

By examining the light from distant sources that has traversed a void, as noted earlier, astronomers often find dark lines in the electromagnetic spectrum, which are due to atoms absorbing some of this light as it passed through eons ago. The light from distant quasars, for example, reveals dark absorption lines caused by hydrogen gas and sometimes carbon, oxygen, silicon, and iron. These elements seem to turn up in gigantic clouds as big as galaxies. So far, the amount of matter found in these intergalactic clouds is about as much as the luminous matter already observed in galaxies, but it is still not enough to account for all the missing baryons.

Whether these clouds are stillborn galaxies that will never shine or young galaxies that should eventually form stars remains an open question. The 1989 observation of a relatively nearby rotating cloud of hydrogen gas on the verge of collapse (see Chapter 6) favors the latter interpretation. And those clouds that contain heavy elements clearly *must* have experienced some kind of stellar nucleosynthesis and had subsequent supernovae to eject the heavy material back into space. But are the stars responsible just too dim to see? Or is something else, something even more curious, going on?

In large clusters and superclusters, x-rays are often observed from intergalactic gas being heated at their high-density centers. These huge ensembles of galaxies provide a unique laboratory, in fact, for studying the nature of the intergalactic medium. Although the amount of gas found so far in the early surveys of x-ray sources cannot account for all the dark baryons, more sensitive searches now being planned may find additional quantities. In particular, ROSAT (short for ROentgen SATellite) was launched in 1990 to search the heavens for x-rays from extragalactic sources.

Another way dark baryons might eventually show up in intergalactic voids is in the form of low-mass objects such as brown dwarfs, which shine with luminosities too low to have been detected yet. Once the Hubble Space Telescope has been repaired, these objects may finally turn up. Or perhaps they will be detected via their infrared radiation, using future satellite-borne detectors sensitive to this part of the spectrum.

If there were a clump of dark matter lurking in a void, baryonic or not, it should act as a gravitational lens — bending the light rays of more distant objects as they traverse the void en route to Earth. Such an effect might occur, for example, if a large galaxy sits between Earth and a distant quasar. The quasar light is bent and focused by the gravitational field of the galaxy, so that images of the quasar appear in two or three places near

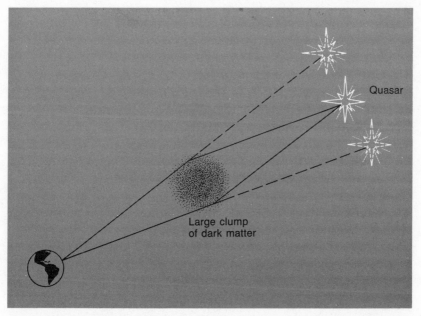

Gravitational lensing of light from a quasar by a clump of dark matter. The light is bent slightly as it passes through and around the clump, generating multiple images of a single bright object.

the galaxy, rather than as a single point of light. Astronomers have already identified several such identical twins or triplets —plus arcs of light thought to occur from the same gravitational lensing effect.

Large clumps of dark matter in a void would affect quasar light in a similar fashion, giving us an indirect way to detect them. If we encounter arcs of light or identical double or triple images but find no galaxy close by to induce these effects, it would be a telling indication that something dark and massive is lurking in the neighborhood.

One must be careful, however, to prove that there really *are* multiple images of the same object—rather than two or three adjacent objects that merely look alike. Such a pair was discovered in 1986, for example, that was claimed to be caused by gravitational lensing. For a while it caused great excitement. But when the initially invisible portions of both electromagnetic spectra were subsequently examined, the pair was found to be actually two separate objects.

In a similar fashion, black holes or brown dwarfs in the Milky Way halo can bend the light of extragalactic sources as it passes by en route to Earth. If they existed in sufficient numbers, we might observe a slight "twinkling," or quivering, of faraway sources as these invisible "lenses" moved past our line of sight. Nearby dwarf galaxies called the Magellanic Clouds, which are observable from the southern hemisphere, provide an excellent series of sources whose light should pass through the dark halo of our galaxy. Two teams of scientists, one from the Lawrence Livermore Laboratory and the other from the Saclay Laboratory near Paris, are instrumenting telescopes in Chile specifically to look at the Magellanic Clouds for this twinkling. Any such gravitational quivering would have to be distinguished, however, from the twinkling that naturally occurs as light encounters density and temperature fluctuations in the Earth's atmosphere.

Resolving the location of the dark baryons has important implications for hot and cold dark-matter models of the Universe. If galactic halos are indeed made of baryonic dark matter, there is no need for cold dark matter to do the job, and hot dark matter would consequently be favored to make up the majority of the Universe. If these baryons are instead found in stillborn galaxies or other clouds of normal matter lying in the intergalactic voids, then cold dark matter would be needed for the halos.

———————————— • ● • ————————————

Another important goal of astronomers is to confirm and extend the observations of large-scale features in the Universe. These kinds of studies are only in their preliminary stages. Far more accurate and extensive data should eventually become available in this decade.

Several groups are busy checking the observations of high peculiar velocities in our local galactic neighborhood—the surrounding 300 million light-years or so that appear to be surging along with the Milky Way at a speed of about 600 kilometers per second. Making such a measurement is difficult, remember, because astronomers have to determine the distances to other galaxies accurately. This is no small feat. But careful work must be done here, because high peculiar velocities, if verified, spell disaster for many scenarios of large-scale structure formation.

The way clusters bunch together requires further study, too, because this is one of the principal arguments favoring topological defects to be the seeds for structure formation. Here the main problem is statistical in nature. At the huge distances characteristic of clusters, there are just not that many objects catalogued yet. The conclusions about cluster correlations are still tentative and may be due to projection effects. We need much more extensive, unbiased samples of large-scale structures, so that we can pick out clusters and determine how strongly correlated they are with each other. Satellite-based x-ray observations may be one good way of generating such unbiased catalogues.

Present studies of correlations rely primarily on one major catalogue of galaxies and clusters, the Abell catalogue, which was generated more than 20 years ago by the late UCLA astronomer George Abell. This catalogue was *not* developed for the types of studies now being attempted with it. More and better data are urgently required, including positions and redshifts of clusters at far greater distances than are presently available. We also need more extensive information about objects visible only from the southern hemisphere. A new survey led by George Efstathiou of Oxford aims to obtain redshifts for 10,000 galaxies in southern skies. Photographic plates are scanned *automatically* for galaxies and clusters, thereby eliminating observer bias.

Large telescopes dedicated to the task of mapping out the structure of the Universe would help resolve these problems.

But given the cost of building one, astronomers have until recently been reluctant to dedicate a major telescope to a single task—even one of such immense importance to cosmology. More typically, telescopes are used for a different task every night, making it difficult to carry out a large-scale survey. As astronomers realize what these systematic surveys may reveal about fundamental physics, however, they are beginning to advocate such a single-task approach for use of their telescopes.

The University of Chicago, Princeton University, and the Institute for Advanced Study are now building a 2.5-meter telescope whose sole purpose is the mapping of galaxies. This telescope will use fiber optics to obtain the spectra of many galaxies simultaneously in any given observation. Because the amount of data to be gathered in this project is enormous, comparable to that of a high-energy physics experiment, particle physicists from Fermilab will be involved in the data analysis effort. In a 5-year period, these scientists expect to measure redshifts for over 1 million galaxies, compared with the present total from all sources of about 10,000. With that many galaxies, large-scale structures can be accurately mapped, and the statistical significance of various arguments will no longer be in doubt. If structures with sizes in the hundreds of millions of light-years are real and ubiquitous, we will be forced to reject random ripples in the primordial sea of matter as the seeds of structure formation.

When the problems in its optics are corrected, the Hubble Space Telescope should begin opening a new era in astronomy. High in orbit above the Earth's atmosphere, it will detect objects of given brightness about 10 times farther away than can now be seen with ground-based telescopes. We may even be able to observe galaxies as they appeared during their epoch of formation over 10 billion years ago. Did they form by matter accreting upon a cosmic string? Did explosive shock waves help trigger the collapse? Or did some other unforeseen event occur? Did quasars exist when the Universe was less than 100 million years old? Were they ubiquitous at an age of 500 million years?

The Hubble Telescope may eventually provide the answers to these kinds of questions. But it, too, is a single facility that will

The Hubble Space Telescope, which was carried into Earth orbit by the Space Shuttle in April 1990. This 2.4-meter telescope should eventually provide a unique glimpse of processes occurring during the epoch of galaxy formation.

have many prospective users competing for the limited time available. By necessity, systematic surveys on this telescope will probably give way to searches for celestial curiosities.

In the late 1990s, NASA plans to launch an instrument called the Advanced X-ray Astronomical Facility (AXAF), to be used for x-ray astronomy in the same manner as the Hubble Telescope will be used for optical astronomy. X-rays are generated by atoms heated to high temperature — millions of degrees — that occur under conditions of extreme violence. If explosions of any kind are involved in galaxy formation, for example, there should be lots of hot gases emitting x-rays, which AXAF will be able to detect. With it, we should be able to tell whether or not explosions caused galaxy formation and thereby determine if gravity got a helping hand from shock waves.

As mentioned earlier, large clusters of galaxies are also known to have hot, x-ray-emitting gas in them, a possible locus of the dark baryons. But how ubiquitous is this hot gas? We do not know yet. AXAF will be able to make the extensive surveys that should tell us whether or not such hot gas can ever account for all the dark baryons.

In addition to these space facilities, there is a new generation of large, ground-based telescopes coming along. For many decades the largest working telescope was the Hale Telescope on Mount Palomar, near San Diego, with a mirror 5 meters in diameter. A 6-meter device built by the Soviet Union at Zelenchukskaya, in the Caucasus Mountains, never performed up to expectations. Now several different groups are planning or building telescopes with mirrors 8, 10, and even 16 meters across — sometimes in clusters of two or more connected optically or electronically. With far more light-gathering capability, this new generation of telescopes will be able to pick up much fainter objects than are observable today. Like the Hubble Telescope, they may provide windows on the epoch of galaxy formation or allow astronomers to discover new objects at extremely high redshifts.

One of these new telescopes is the Keck Telescope being built by CalTech and the University of California. With a huge, segmented mirror 10 meters across, it has been situated atop Hawaii's Mauna Kea in order to minimize the atmospheric distortion that can limit its resolution — its ability to distinguish objects of small apparent dimension. A second version of this unique telescope is now in the planning stages.

Another large telescope is being planned by a consortium that includes the University of Arizona, Ohio State University, and the Osservatorio Astrofisica di Arcetri in Italy. Called the Columbus Project, it will employ twin 8-meter mirrors placed 6 meters apart, giving it the resolution of a 22-meter telescope ($8 + 6 + 8 = 22$) and the light-gathering ability of an 11-meter device. (The combined area of two 8-meter disks is equal to that of a single 11-meter disk.)

Finally, a consortium of Western European nations including

A model of the Very Large Telescope. This array of four 8-meter telescopes is being built by a European consortium in the mountains of northern Chile.

France, Italy, and West Germany has decided to build the Very Large Telescope in the mountains of northern Chile. A linear array of four 8-meter mirrors that can be used independently or linked by optical and computer means, it will have the combined light-gathering ability of a mirror 16 meters across, making it easily the world's largest telescope.

Once these large telescopes are built, they should have a salutary effect upon the use of the smaller devices. Astronomers who concentrate on weird objects at high redshifts will probably devote more time to the big telescopes, freeing the smaller devices for extensive galaxy surveys of the kind that discovered the foamy large-scale structure of the Universe. We may finally escape from the current state of affairs, in which telescope time is divided among a huge number of unrelated projects, and instead begin to obtain a more coherent picture of the Universe as a whole.

A 4-meter telescope totally dedicated to determining redshifts of galaxies in the southern hemisphere would be a great boon to cosmology. Together with the Chicago–Princeton 2.5 meter telescope now underway, it would help us generate a three-dimensional map of the Universe out to far greater depths than has yet been accomplished. Such a map would determine how ubiquitous the foamy structure really is, and whether it evolves with time. Are the voids really bubbles? Does the bubble size remain the same? Is the structure truly sponge-like? Or sheet-like? These are the kinds of questions we will be able to answer as more and bigger telescopes are dedicated to galaxy surveys at greater and greater distances.

New surveys will also permit more detailed checks to be made of the measurements of Loh and Spillar (see Chapter 8), who employed the apparent density of galaxies to conclude that space was indeed flat — consistent with $\Omega = 1$. Their results are questioned because galactic evolution may have affected their ability to pick out galaxies at the greatest distances, thereby biasing their data. Systematic galaxy surveys should allow us to determine such effects, if they exist, and to make any corrections needed. That way we can eventually make an unambiguous measurement of the parameter Ω, to serve as a check on theoretical prejudices.

———————————— • ● • ————————————

Extensive galaxy surveys will provide a crucial test of cosmic strings and other models of structure formation based on topological defects. By mapping out the large-scale structure of the Universe, traditional optical astronomers (whose field of study dates all the way back to Galileo) can now use their telescopes to explore unified theories of elementary particles. The macrocosm may yield important insights about the microcosm that could never be obtained even at the most powerful atom-smashers.

Another possible way to detect cosmic strings and other defects would be by their action as gravitational lenses. A length of cosmic string should serve as a long lens — inducing a series of double images. Leonard Cowie at the University of Hawaii recently discovered a group of objects that appear to contain many such identical pairs. Some cosmologists have suggested that these pairs reveal the existence of a cosmic string, but others think the pairs are merely a series of binary galaxies like M31 and the Milky Way.

Cosmic strings and other defects would also induce tiny variations in the temperature of the microwave background radiation. As shown by Albert Stebbins of Fermilab, the temperature of the radiation just above a string would be slightly different from that just below it. As a physicist scanned across the string, he or she would notice a very small "step" — a difference of about ten parts per million. The same small step would occur all along the string. Although such precision is currently at the limits of detection capability, new and better instruments will soon make this approach a good way to hunt for topological defects.

Detailed searches for ripples in the cosmic background radiation are being done by the COBE satellite as well as by new ground-based detectors at the south pole and elsewhere on Earth. These experiments will eventually push the accuracy of such measurements to the level of a few parts per million. If we do not find ripples in the background radiation at that level, we may be forced to consider late-time phase transitions (see Chapter 10) as the source of large-scale structure. If ripples *are* discovered, we would have to determine whether their pattern was random or correlated. If the latter, it would represent evidence for some kind of cosmic string or other topological defect. These kinds of studies may even be able to detect the primordial cosmic seeds directly.

Other evidence for cosmic strings may come from their effect on the radiation from pulsars, which are neutron stars rotating very rapidly. The best "clocks" known, as good as or better than atomic clocks, some pulsars emit intense bursts of radio waves at *extremely* regular intervals on the order of a few milliseconds. As loops of cosmic string vibrate, they would emit gravitational radiation that alters very slightly (about ten parts per million) the distance from a pulsar to the Earth. The apparent period of the pulsar—the time between two successive bursts—would fluctuate due to this "warping" of the intervening space. The absence of any such variations in measurements made so far has already eliminated many possible models of cosmic strings. As their precision improves, scientists should encounter an intrinsic level of fluctuations in the observed pulsar timing if any such model is correct.

Although explosions of superconducting cosmic strings are now known not to have generated large-scale structures, they may nonetheless still exist and produce observable effects. In particular, evidence for them may turn up in ultrahigh-energy cosmic rays. As the currents of charged particles circulating inside shrinking loops of this string become too intense, the loops break down and disgorge very high-mass, high-energy particles. Such expiring loops should also be a major source of ultrahigh-energy corpuscles, some of which would eventually impinge upon Earth.

Indeed there *are* cosmic rays striking the Earth at extremely high energies—up to almost 100 billion GeV, about 100 billion times the mass–energy of a single proton. That's a pretty powerful particle. For comparison, the highest energy protons generated by an earthbound particle accelerator, the Tevatron at Fermilab, carry energies only up to 900 GeV. The origin of these ultrapowerful cosmic rays has not been firmly established. But superconducting cosmic strings may in fact be the source, suggests physicist Chris Hill of Fermilab. Observations of these cosmic rays, therefore, help to constrain the various possible models and limit their arbitrariness.

The Fly's Eye Observatory at the Dugway Proving Grounds in Utah is a collection of electronic devices for detecting ultrahigh-energy cosmic rays that trigger enormous showers containing many subatomic particles when they strike the upper portions of the Earth's atmosphere. As these secondary particles zoom down through the lower atmosphere, they produce faint flashes of light that can actually be detected by the individual elements in the Fly's Eye.

Built and operated by the University of Utah, the Fly's Eye Detector is much larger than any previous cosmic-ray detector array. It surveys a thousand square kilometers—about 400 square miles—of the sky. In only a few years of operation, it has more than doubled the ultrahigh-energy cosmic-ray events witnessed by all the other detectors in the world combined.

Now other groups, including one from the University of Chicago led by James Cronin and another from the University of Michigan led by John van der Velde, are joining this project. These physicists are positioning additional arrays of detectors on the land surrounding the Fly's Eye and in earth beneath it. The hundreds of devices added by the Chicago group will provide

A portion of the Fly's Eye Detector in Utah. With multiple mirrors aimed at adjacent portions of the sky, this device detects ultrahigh-energy cosmic rays as they strike the upper atmosphere.

far more information about the particle showers, including their energy and direction; those from the Michigan team will help identify penetrating particles known as muons (see Chapter 2) in the showers. Given time and the combined capacities of all these detectors, the Fly's Eye Observatory should eventually be able to tell whether or not ultrahigh-energy cosmic rays actually began with the death throes of superconducting cosmic strings.

This array of particle detectors might even detect ultrahigh-energy neutrinos from space. If they pass through the Earth and interact with matter just before surfacing, they would initiate upward-moving showers of particles that could be detected in the array. In a similar fashion, large underground detectors have already been used to search for upward-going muons that would result if high-energy neutrinos from space interacted just before reaching them. An exciting new field of study, using the techniques of particle physics to search for the high-energy remnants of astrophysical objects, has recently begun to do serious research.

———————————— • ● • ————————————

A new breed of physicist has recently begun the difficult search for dark matter itself—or at least for direct evidence that this shady stuff truly exists. If the halo surrounding the Milky Way is indeed composed of dark, nonbaryonic particles, these scientists reason, why not try to detect them? These dim corpuscles may not be altogether obvious, but they should lurk everywhere, permeating the very space we inhabit and the air we breathe. Perhaps ultrasensitive measurements of one kind or another might be able to discover faint traces of these timid ghosts. In 1989, the National Science Foundation began funding a new Science and Technology Center at the University of California at Berkeley devoted primarily to the direct search for dark matter.

Those physicists enamored of cold dark matter have a special motivation to seek it in the local halo, because it *should* have clumped here with baryonic matter. Hot dark matter, on the other hand, would be far more thinly distributed. Because it is faster moving, it would aggregate on much larger distance scales, which would make it more difficult to detect.

These cold, dark particles have become known by the general acronym "WIMP," for Weakly Interacting Massive Particle, a waggish name coined by physicist Michael Turner at the University of Chicago. All WIMPs have some amount of mass, and they can interact only feebly with ordinary matter—or else we would have already observed them. In the late 1980s, WIMP searches became the order of the day among physicists interested in dark matter. A modern-day scientific gold rush has begun.

If they are massive enough to gather in the local halo, WIMPs should also collect in very high density objects such as the Earth and Sun. Although their interaction with protons and neutrons is exceedingly feeble, chance encounters will still happen occasionally when a WIMP passes through one of these orbs. The WIMP surrenders some energy as it ricochets—and slows down a bit. After several such encounters, it may lose enough speed and energy to become gravitationally bound. Greater concentrations of WIMPs would occur near the Earth and Sun as these orbs continually sweep through the halo, picking up new adherents. So the best place to look for dark matter may be right under our very own feet!

For awhile, it seemed there might be indirect evidence for WIMPs collecting inside the core of the Sun, as suggested by John Faulkner of the University of California at Santa Cruz and William Press of Harvard. If WIMPS had the right mass, these scientists claimed that they would help cool the Sun's core slightly by transporting energy to its outer reaches. This loss would lower the temperature of the Sun's core and slow the rate of thermonuclear reactions occurring there. Because these reactions generate neutrinos that are able to escape from the core, this chain of events would reduce the flux of neutrinos being emitted from the Sun.

Such an effect may already have been witnessed by Raymond Davis of Brookhaven National Laboratory. For more than two decades beginning in the 1960s, he has patiently studied solar neutrinos using a huge tank of perchloroethylene (a common cleaning fluid) situated deep in the Homestake gold mine beneath the Black Hills of South Dakota. Every two days or so, a neutrino from the Sun struck a chlorine atom in the tank, changing it into an argon atom, which Davis later extracted and detected by very sensitive methods able to count individual atoms. The rate of solar neutrinos he observed is only about one-third of what is expected in standard models of the Sun's core. More recent and accurate studies of the solar neutrino flux, done by a group of scientists using a huge underground detector in Japan's Kamiokande mine, have confirmed these findings, putting the measured rate at about 40 percent of that expected. Such a deficit of solar neutrinos, Faulkner and Press claimed, can be readily explained by WIMPs with masses between 2 and 10 GeV that spirit energy away from the core, cooling it by 10 percent. (Other theorists, of course, offer other explanations.)

But WIMPs collecting in the core of the Sun (and Earth) should have had another important effect. In such enhanced concentrations, WIMPs would often encounter their antiparticles (the antiWIMPs), too. The couple would annihilate one another, occasionally producing very high energy neutrinos that would escape the Sun with energies greater than 1 GeV. These neutrinos should leave very distinctive signatures in the large underground detectors buried around the world to watch for proton decay (see Chapter 5). In addition to the Kamiokande detector in Japan, there is the IMB detector in a salt mine near Cleveland, Ohio, built by a collaboration from the University of California at Irvine, the University of Michigan, and Brookhaven National Laboratory. These two enormous tanks of water were able to record about 20 neutrinos with much lower energies emanating from the 1987 supernova. But they have so far failed to detect any high-energy neutrinos that cannot be explained by other, more conventional means.

The fact that no high-energy neutrinos have been witnessed coming from the Sun casts serious doubt on Press and Faulkner's proposition. The lack of any such distinctive events also seems to indicate that several of the more popular cold dark-matter candidates, like photinos and other neutral SUSY particles, must be more massive than 10 GeV if they are to account for the dark matter in the halo of our galaxy. The explanation of the solar neutrino discrepancy probably lies elsewhere (perhaps in the possibility that neutrinos have a very tiny mass).

A scuba diver inspecting the interior of the IMB detector, which contains over 8000 tons of ultrapure water. Lining the walls are thousands of photomultiplier tubes sensitive to minute flashes of light that charged particles make as they pass through the water.

Other evidence for heavy WIMPs — photinos for example — might be the detection of high-energy photons in cosmic rays. If a photino met its antiparticle in the halo, they would annihilate each other and occasionally produce a pair of energetic photons. Or they might produce proton–antiproton pairs, too. Either alternative would lead to a distinctive, monoenergetic signature in cosmic rays. The lack of any such observation so far puts stringent limits on these possibilities.

Many physicists around the world are now attempting to detect the occurrence of a single WIMP as it dislodges a proton or neutron in ordinary matter. To have any chance at all of seeing such a rare encounter, they have to put enough matter in the path of the WIMP and still be able to detect the incredibly tiny energy deposit that occurs when a single proton or neutron recoils. Devices able to accomplish such a feat have to be extremely well shielded from accidental cosmic rays and background radioactivity that can easily masquerade as a spurious signal. And they usually must be cooled to very low temperatures close to absolute zero, because the minuscule energy deposits expected of a true WIMP would otherwise be swamped by the normal thermal vibrations of the atoms themselves. It is a challenging task indeed, but an ever-growing number of physicists is game to try.

One of the best results thus far in such laboratory searches for dark matter was reported in 1990 by David Caldwell and his coworkers from the University of California at Santa Barbara, Lawrence Berkeley Laboratory, and Stanford University. These physicists were using extremely pure crystals of germanium to search for certain rare forms of nuclear decay. They realized that their detectors are sensitive enough, however, to record the effects of electrons that would be knocked about by protons dislodged after certain kinds of WIMPS collided with them. From the absence of any such effects, they concluded that these kinds of WIMPs had to have masses less than 15 GeV, or they must interact with matter far more weakly than neutrinos do. These results also make it particularly unlikely for WIMPs to exist with the properties needed by Faulker and Press to explain the deficit of solar neutrinos. And when combined with the lower limits on the mass of any additional neutrino (beyond the three kinds already known) obtained from LEP and SLC experiments on the Z particle, this experiment can exclude very heavy neutrinos as a viable dark-matter candidate.

Another direct detection method, originally proposed by the Polish physicist André Drukier and developed further by several other European scientists, involves the use of many, many tiny beads of superconducting material. When the beads are cooled to sufficiently low temperatures, they become superconductors. In addition to having no resistance at all to the passage of an electric current, superconductors do not allow external magnetic fields to enter them. Now suppose a WIMP were to penetrate a bead and dislodge one of its protons or neutrons. This corpuscle might deposit enough energy to warm the tiny bead to the point where it was no longer superconducting. The bead "goes normal" in the usual jargon; an external magnetic field can once again get inside. Such a distinctive change of state can be detected, even in a single tiny bead, by what is called a "superconducting quantum-interference device."

Other, similar methods of direct WIMP detection are being developed. Most of them also involve hunting for tiny energy deposits in a small amount of supercold matter. All face the same serious problem that thermal fluctuations, accidental cosmic rays, background radioactivity, or even the innocent rumble of a passing truck can induce similar effects without the intercession of any WIMPs whatsoever. No convincing evidence of a WIMP–matter encounter has yet been claimed, but this kind of research has only recently begun. This rapidly evolving field may produce important results in the near future.

The methods we have mentioned so far are best suited for *heavy* forms of cold dark matter, such as photinos or very heavy neutrinos, with masses greater than 2 GeV. Laboratory detection schemes for axions, on the other hand, involve different but equally novel techniques. If responsible for the surrounding halo, these wispy, ethereal particles would be absolutely ubiquitous — far more common, even, than neutrinos or light

244 · *The Shadows of Creation*

(trillions per cubic inch, by certain estimates). But they would be so light and so slow-moving that their impacts upon normal matter would be absolutely imperceptible, and far beyond all hope of detection. Thus, we have to employ measurement techniques that take advantage of their vast numbers.

A common feature of axions, in their many manifestations, is their propensity to disintegrate into two photons of equal energy. The heavier the axion, the more quickly this disappearing act has to occur. If axions carried masses of 8 eV, for instance, and if the halo of a cluster were composed of them, it would be anything but dark. Instead, it would glow with a soft blue light —4-eV photons emitted in all directions. The fact that such a glow has not been detected, despite extensive searches by Michael Turner and colleagues, means we can safely rule out axions with masses in the neighborhood of 2 to 10 eV.

Other astrophysical arguments can be used to constrain the possible masses of axions. If they weighed in at around 1000 eV, for example, the Sun would cool too rapidly by axion emission; the fact that the Sun rises every day means there are no axions with such a mass. Helium burning in red giant stars would never even begin if there were axions with masses above about 0.01 eV. Since it *does* in fact occur, in such obvious red giants as the star Antares in the constellation Scorpio, any such axion must be lighter than this.

When first proposed in the early 1980s "invisible axions" (see Chapter 9) were thought to be absolutely undetectable: they could never be observed. These wispy ghosts had to be so light and so ethereal that their disintegration into photons would take many, many times the age of the Universe. And any photons that did result would be virtually impossible to spot. But since those early years, several interesting laboratory detection schemes have been proposed—and certain venturesome physicists have already begun to attempt such a measurement.

In 1983, theorist Pierre Sikivie of the University of Florida noted that "invisible" axions might be intentionally converted into microwave photons in the presence of an extremely strong magnetic field. If this field was applied to a copper cavity cooled almost to absolute zero, so that thermal vibrations did not swamp true signals, one might be able to detect axions by the spontaneous appearance of excess energy in the cavity when it was tuned to exactly the right microwave frequency. Any axions permeating the cavity would be converted there into microwaves, boosting its energy yield.

The first experiment along these lines, performed by a collaboration of Brookhaven, Fermilab, and University of Rochester physicists led by Adrian Melissinos, was reported in 1987. To obtain an extremely strong magnetic field, they used a large superconducting magnet available at the Brookhaven National Laboratory on Long Island. Their quarry was an invisible axion with the unimaginably tiny mass of 0.000005 eV — ten-*trillionths* of the mass of the electron itself! These intrepid scientists came away empty-handed, but the sensitivity of their equipment was about 100 times too low to have ever detected such an axion anyway. Undaunted, they are upgrading the apparatus and planning to continue the hunt. A new team, led by Karl Van Bibber of Lawrence Livermore Laboratory (and including Sikivie and Turner), is now attempting to build a much larger cavity with enough sensitivity to detect axions in the mass range that can yield $\Omega = 1$.

Even the spectacular 1987 supernova, SN1987A, can supply information about axions. The fact that electron neutrinos from this stellar paroxysm were witnessed on Earth with the expected numbers and energies can only mean that they, and not axions, must have provided the principal means by which energy escaped from the collapsing core of this dying star. The core was "cooled" by neutrino emission, that is, not by axion emission.

Several independent teams of physicists have used this observation to put the most stringent limits thus far obtained on the possible mass of any invisible axion. It must be less than 0.001 eV, according to one group (which includes David Schramm). Only a narrow range of axion masses now remains viable — those between about 0.00001 and 0.001 eV. This is just the range of axion masses that can yield $\Omega = 1$, so there is great interest in the next generation of axion searches. The

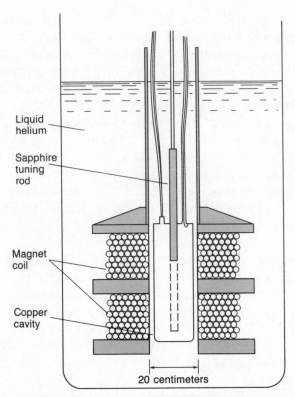

Liquid helium

Sapphire tuning rod

Magnet coil

Copper cavity

20 centimeters

Cutaway view of the axion detector built by a collaboration of Brookhaven, Fermilab, and University of Rochester scientists. Axions decaying inside the copper cavity would make it resonate at microwave frequencies; the trick is to detect these faint oscillations.

supernova constraints on the axion mass represent one of the most impressive results thus far obtained in the burgeoning field of elementary-particle astrophysics.

If the dark matter happens to be hot, then the likeliest candidate is a massive neutrino, with a mass somewhere between 20 and 30 eV. At least we *know* that neutrinos exist in sufficient quantities, which is far more than we can say for all the other, more exotic corpuscles currently being advocated. Although the electron neutrino has already been excluded from contention (because its mass is known from laboratory measurements to be less than 10 eV), the muon neutrino and tau neutrino still remain viable candidates. Our esthetic and theoretical prejudices favor this possibility, too, because we naturally expect that the neutrino partners of the heavier leptons would be more massive themselves.

In a number of fairly sensitive experiments, the masses of the muon and tau neutrinos have been measured to be less than 250,000 eV and 35 MeV, respectively. Cosmology requires that they both weigh in at less than 30 eV, however, because heavier neutrinos would have affected the dynamics of the early Universe in ways that are inconsistent with what is observed. So today's experimental limits are still very far from the cosmologically interesting range of neutrino masses.

Direct laboratory measurements of the masses of these two neutrinos are almost impossible, however, because they are invisible decay products of far heavier particles — the muon and tau particle. We would have to measure many, many decays with extraordinarily high precision, far beyond current experimental capabilities, to be able to conclude that either or both of these neutrino masses are in the 20 to 30 eV range. All we can offer today are crude upper limits, those given above, which will come down only gradually with time.

One hope for measuring the masses of the muon and tau neutrinos (if either falls in this range of cosmologically interesting masses) would be from the detection of them in a nearby supernova. Remember that the 1987 supernova, which was located in the Large Magellanic Cloud about 170,000 light-years away, produced enough electron neutrino events in two underground detectors for physicists to establish an upper limit of 25 eV on its mass (see Chapter 9). If such a supernova occurred in the Milky Way, we might observe *thousands* of neutrinos in underground detectors and begin to distinguish one kind from another.

The Sudbury Neutrino Observatory, an underground detector scheduled for construction in the Canadian province of Ontario during the early 1990s, should be particularly sensitive to tau neutrinos because it will use heavy water (based on deuterium instead of hydrogen, D_2O instead of H_2O) as its active element. Although this is a long shot at best, we may get lucky. Perhaps a supernova occurring in our vicinity before the end of the century can tell us whether the tau neutrino is more massive than the others.

Indirect measurements of a small but finite neutrino mass might still be possible, too, because of a quantum mechanical phenomenon known as "neutrino mixing." If neutrinos have masses, even exceedingly tiny masses equal to a fraction of an electron volt, then the various species would "mix" with one another. A physical neutrino prepared in the laboratory, that is, would actually be a *combination* of two or three different species. The neutrino emerging from the decay of a muon, for example, would spend most of its time as a muon neutrino but tiny fractions of its time as an electron or tau neutrino. By measuring the extent of such mixing, one might be able to gauge the masses of the various species.

In 1980, a preliminary result that seemed to indicate such mixing was true led to great excitement about the possibility that neutrinos indeed had a small mass (see Chapter 9). But that result was not confirmed, and since then experimenters have found no clear evidence for neutrino mixing. That, however, does not dissuade them from continuing the hunt. Experiments at Fermilab, for example, may soon push the search for tau neutrino mixing into the cosmologically interesting 20 to 30 eV range.

Neutrino mixing may also be responsible for the solar neutrino deficit witnessed in the Homestake and Kamiokande experiments mentioned earlier. Most of the electron neutrinos produced by nuclear reactions deep in the Sun's core may be converting into muon or tau neutrinos as they speed through its outer layers. But the latter two neutrino types would simply not register in these underground detectors, thereby leading to the deficit observed. In order for such "matter oscillations" to occur, however, one or more of these three neutrinos has to have a small but finite mass — on the order of 0.001 to 0.01 eV.

A similar deficit was reported in 1990 by a Soviet-American experiment under the Caucasus Mountains between the Black and Caspian Seas in the USSR. This was the first experiment to use large quantities (30 tons) of the element gallium as the active detector material, which makes it sensitive to the great bulk of lower-energy electron neutrinos that are produced in the Sun's core (unlike the Homestake and Kamiokande detectors, which see only the rare high-energy neutrinos). Although these results are preliminary, we may finally be obtaining the first concrete evidence for a non-zero neutrino mass, which would be an extremely important discovery.

If true, the mass of the electron and muon neutrinos would still be far too small — less than 0.01 eV — to close the Universe. But in the favorite theory for the origin of neutrino masses, known as the "see-saw mechanism," the tau neutrino could then have a mass of around 30 eV. This is just about the right neutrino mass needed to make $\Omega = 1$.

———————————— • ● • ————————————

Traditional particle physics experiments at the major accelerators and colliders around the globe are another part of the ongoing hunt for cold dark-matter candidates. Most frequently sought are the supersymmetric, or SUSY, particles such as the photino or gluino. The lightest member of the SUSY family, you remember, should be massive and stable. Along with a few other desirable properties, that makes it an excellent prospect for the dark matter in the Universe.

The best kinds of facilities to use for such particle searches are the colliding-beam machines, in which two beams of high-energy particles traveling in opposite directions at close to the speed of light clash with one another. Generally one beam contains stable particles of ordinary matter—electrons or protons—and the other beam contains their antiparticles—positrons or antiprotons. When a particle and its antiparticle meet at the clashpoint, they annihilate each other. All the tremendous energy bound up in their motion suddenly becomes available to make new, exotic, and *heavy* particles.

Experimenters surround the clashpoint with concentric layers of particle detectors, in an attempt to record the trajectories, energies, and other identifying characteristics of subatomic particles emerging from the violent smash-ups. Using computer-aided reconstructions of these visible remnants, physicists try to determine what *else* might have been created in the debris—and what might have escaped detection altogether. A neutrino, for example, leaves no visible trace whatsoever. But its occurrence can still be determined indirectly, by proving that an amount of energy is missing in a specific portion of the surrounding detector.

Collision events with such a "missing-energy" signature might also signal the creation of a stable, neutral SUSY particle such as the photino. Imperceptible by normal techniques, it would leave the clashpoint and pass through the surrounding detector layers completely unseen, its only trace being an unexplained amount of energy that did not turn up as expected in these detectors. Whatever carried away the missing energy would have had to be something *dark*.

Such events with a missing-energy signature were indeed found, in 1984, by a collaboration of European physicists, led by Carlo Rubbia, who were working with a proton–antiproton collider at CERN. For a time there was great excitement that perhaps SUSY particles had actually been discovered, with masses less than 50 GeV. But further analysis and experiments, both by Rubbia's team and other groups, proved that such an interpretation was not valid. The missing energy could be ex-

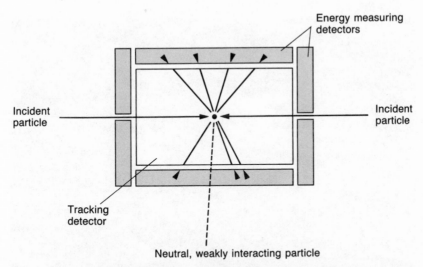

An event with "missing energy," as might be observed in a large detector that surrounds the point where two particle beams clash. A neutral particle that interacts very weakly with matter would escape without leaving a trace, and more energy would appear to be deposited on one side of the detector than the other.

plained as due to ordinary processes, such as the decay of a tau lepton into an (invisible) tau neutrino and other debris. More recent experiments at CERN indicate that the mass of a photino, if it exists, has to be greater than 15 times the proton mass.

The search continues, however, in the energy domain opened by the Tevatron Collider at Fermilab. On this machine protons and antiprotons can be accelerated to energies close to 1 *trillion* electron volts, or 1 TeV—almost 1000 times the energy contained in their intrinsic masses. The protons and antiprotons collide with one another inside an enormous detector known as the Collider Detector at Fermilab (CDF). With so much energy concentrated inside a tiny region (about the size of a proton), it is easy to create new particles with masses well beyond that of the Z particle, which is the heaviest elementary particle definitely known to exist. Using this collider and detector, scientists have already established that squarks and gluinos must have

The Collider Detector at Fermilab, or CDF. Surrounding the point where high-energy protons and antiprotons collide, this detector is being used by physicists to search for the telltale traces of extremely massive, exotic particles.

masses greater than 100 GeV. Perhaps the photino will eventually show up here, in the guise of missing-energy events.

Meanwhile, a new generation of electron–positron colliders has begun detailed studies of the massive Z particle. The Stanford Linear Collider, or SLC, speeds electrons and positrons to energies of 50 GeV and crashes them together in a single pass. It is designed to produce large numbers of this particle, at a mass of 91 GeV. At CERN, the Large Electron–Positron collider, known as LEP, began operations in 1989 with the same purpose. By 1990 this 27-kilometer storage ring had produced almost 800,000 Z particles.

In their first round of measurements, the SLC and LEP have proved that there are only three different kinds of light neutrinos — and no more. By confirming a prediction based on Big-Bang nucleosynthesis, physicists using these colliders have verified a key cosmological argument and placed a firm cap on the complexity of the Universe. They also helped kill heavy neutrinos as a viable candidate for cold dark matter. A fourth kind of neutrino with a mass less than about 45 GeV is ruled out by the LEP and SLC measurements, and most higher masses have already been excluded by the underground germanium experiment of Caldwell and coworkers.

The 1990 LEP measurements go even further and restrict the masses and interaction strengths of any *other* kinds of WIMPs that might make a major contribution to the mass of the Universe. A neutral SUSY particle with less than half the mass of the Z particle, for example, would give the Z another way to decay — thereby shortening its lifetime slightly. But the very precise LEP measurements of this lifetime are completely in accord with predictions based on the Standard Model with only three families of quarks and leptons. Nothing else is needed. If there is any WIMP able to occur in decays of the Z, it must be heavier than 15 GeV and its interaction with matter must be at least ten times weaker than that of a neutrino. That's awfully weak. It will be very difficult to detect such a particle directly, assuming it exists, but well-designed underground detectors might still have a chance.

The next major goal for electron–positron colliders is the hunt for the putative Higgs particle. This missing link in the Standard Model is thought to be responsible for the breakdown of the electroweak force and for imbuing particles with mass. This would not be the ultra-Higgs particle discussed at length in Chapter 8, which is centrally involved in the process of inflation. But it is a similar kind of object, and its discovery would have truly immense significance for our comprehension of the Universe. The Higgs particle is widely considered to be the key to the phenomenon of symmetry breaking — a central feature in all unified theories of elementary particle forces.

If the Higgs particle carries a mass less than about 50 GeV, it should eventually turn up at LEP in decays of the Z. But because this will happen very rarely, physicists need to produce many *millions* of Z particles before they could be truly certain of the discovery. By late 1990, CERN physicists had found no evidence for any Higgs particle at LEP, and they were able to conclude that its mass must be greater than about 45 GeV.

If the Higgs mass happens to fall above 50 GeV, scientists working at LEP will still have a shot at finding it. By 1994, this collider should be ready to clash electrons and positrons at energies up to 100 GeV, twice as much as the SLC, allowing LEP to produce pairs of massive W bosons at 80 GeV apiece. These higher energies will also permit extended searches for the Higgs particle, up to a mass of at least 80 GeV.

To examine the mass range far beyond that of the W and Z will require colossal particle accelerators like the Superconducting Supercollider, or SSC, now under construction in Texas. With a pricetag of about $8 billion, this 54-mile ring will speed protons to energies close to 20 TeV in two counter-rotating beams. Proton–proton collisions at total energies up to 40 TeV should be able to create heavy new particles with masses from a few hundred GeV to several TeV. A more modest effort is the 16 TeV Large Hadron Collider, or LHC, being proposed for construction in the existing LEP tunnel at CERN; it could search for particles with masses up to about 1 TeV. If they exist (and have not already been discovered), the Higgs particle and SUSY particles should eventually show up at the LHC and SSC. Or perhaps something else will be found that will require a major modification in the way theorists think about forces. Something has to give, at these truly enormous energies.

By providing our first glimpse of the Higgs field, these colliders may give us crucial information about the very nature of mass itself. That would help physicists to answer deep questions like: Why are things heavy? We might encounter an entity much like the symmetric vacuum of the GUTs epoch, whose remnants may still exist today as topological defects in empty space. These mammoth machines may therefore help us to comprehend the nature of the vacuum itself and to study the fabric of the Universe. This is indeed an exciting prospect.

The LHC and SSC can also help to search for extremely heavy WIMPs — cold dark-matter particles with masses up to several TeV. According to recent theoretical work, this mass level is getting close to the highest possible mass that WIMPs can possess and still be responsible for making Ω equal to 1, the most favored value. Now that masses of 1 to 10 GeV have been eliminated, as mentioned earlier, this TeV mass range is thought to be the most likely place for them to show up. The discovery of such an ultramassive WIMP, responsible for most of the mass of the Universe, might well prove to be more important than the discovery of the Higgs particle itself.

———————— • ● • ————————

Modern science is based on observation and measurement, and cosmology is no exception. The mysteries of dark matter and the structure of the Universe will be resolved not just by thinking and calculating, but also by watching and probing. We need research on many different fronts, using telescopes and satellites, cosmic-ray and underground detectors, plus dark-matter searches in small laboratories and particle searches at gigantic colliders. And we cannot know in advance where the important discoveries will eventually occur. A few completely unanticipated surprises will almost certainly turn up. But by the end of this century, if not sooner, we hope to know what makes up the bulk of the Universe and how this extraordinary matter came to be distributed the way it is.

Further Reading

General Interest Books

Barrow, John D., and Joseph Silk, *The Left Hand of Creation*, Basic Books, 1983.

Bartusiak, Marcia, *Thursday's Universe*, Times Books, 1986.

Close, Frank, Michael Marten, and Christine Sutton, *The Particle Explosion*, Oxford University Press, 1987.

Crease, Robert P., and Charles C. Mann, *The Second Creation*, Macmillan, 1986.

Davies, Paul, *Superforce*, Simon & Schuster, 1984.

Ferris, Timothy, *Coming of Age in the Milky Way*, William Morrow, 1988.

Hawking, Stephen, *A Brief History of Time*, Bantam Books, 1988.

Krauss, Lawrence, *The Fifth Essence*, Basic Books, 1989.

Lederman, Leon, and David N. Schramm, *From Quarks to the Cosmos*, Scientific American Library, 1989.

Pagels, Heinz R., *The Cosmic Code*, Bantam Books, 1983.

Pagels, Heinz R., *Perfect Symmetry*, Simon & Schuster, 1985.

Riordan, Michael, *The Hunting of the Quark*, Simon & Schuster, 1987.

Sutton, Christine, *The Particle Connection*, Simon & Schuster, 1984.

Trefil, James, *The Moment of Creation*, Scribner's, 1983.

Tucker, Wallace, and Karen Tucker. *The Dark Matter*, William Morrow, 1988.

Weinberg, Steven, *The Discovery of Subatomic Particles*, W. H. Freeman and Company, 1990.

Weinberg, Steven, *The First Three Minutes*, Basic Books, 1977.

Wilczek, Frank, and Betsy Devine, *Longing for the Harmonies*, W. W. Norton, 1988.

More Difficult Books

Cahn, Robert N., and Gerson Goldhaber, *The Experimental Foundations of Particle Physics,* Cambridge University Press, 1989.

Carrigan, Richard A., and W. Peter Trower, eds., *Particles and Forces*, W. H. Freeman and Company, 1990.

Carrigan, Richard A., and W. Peter Trower, eds., *Particle Physics in the Cosmos*, W. H. Freeman and Company, 1989.

Mandelbrot, Benoit B., *The Fractal Symmetry of Nature*, W. H. Freeman and Company, 1982.

Pais, Abraham, *Inward Bound*, Oxford University Press, 1986.

Parker, Barry, *Creation*, Plenum Press, 1988.

Reeves, Hubert, *Atoms of Silence*, MIT Press, 1984.

Segrè, Emilio, *From X-Rays to Quarks*, W. H. Freeman and Company, 1980.

Silk, Joseph, *The Big Bang*, revised ed., W. H. Freeman and Company, 1989.

Van Helden, Albert, *Measuring the Universe*, University of Chicago Press, 1985.

Scientific and Technical Books

Abbot, L. F., and So-Young Pi, eds., *Inflationary Cosmology*, World Scientific, 1986.

Audouze, Jean, and Sylvie Vauclair, *An Introduction to Nuclear Astrophysics*, Reidel, 1980.

Bahcall, John N., *Neutrino Astrophysics*, Cambridge University Press, 1989.

Balian, R., Jean Audouze, and David N. Schramm, *Physical Cosmology*, North Holland, 1980.

Borner, Gerhard, *The Early Universe: Fact and Fiction*, Springer Verlag, 1988.

Close, F. E., *An Introduction to Quarks and Partons*, Academic Press, 1979.

Galeotti, Piero, and David N. Schramm, eds., *Dark Matter in the Universe*, Kluwer Academic, 1990.

Kolb, Edward W., and Michael S. Turner, *The Early Universe*, Addison-Wesley, 1990.

Kolb, E. W., M. S. Turner, K. Olive, D. Seckel, and D. Lindley, *Inner Space/Outer Space*, University of Chicago Press, 1986.

Misner, Charles, Kip Thorn, and John A. Wheeler, *Gravitation*, W. H. Freeman and Company, 1973.

Peebles, P. J. E., *The Large-Scale Structure of the Universe*, Princeton University Press, 1980.

Peebles, P. J. E., *Physical Cosmology*, Princeton University Press, 1971.

Perkins, D. H., *Introduction to High-Energy Physics*, Addison-Wesley, 1987.

Rolfs, Claus E., and William S. Rodney, *Caldrons in the Cosmos*, University of Chicago Press, 1988.

Weinberg, Steven, *Gravitation and Cosmology*, John Wiley, 1972.

Zee, Anthony, ed., *Unity of Forces in the Universe*, Vols. I and II, World Scientific, 1982.

Zeldovich, Yacov B., and I. D. Novikov, *Relativistic Astrophysics, Vol. 2: The Structure and Evolution of the Universe*, University of Chicago Press, 1983.

Acknowledgments

*A*ny publication in or about science relies heavily upon the work of others, and this book is no exception. We therefore wish to thank the following scientists for their help and encouragement, and especially for useful conversations about the content of these pages: Charles Alcock, Jean Audouze, John Bahcall, Edmund Bertschinger, George Blumenthal, Richard Bond, David Caldwell, Marc Davis, Alan Dressler, John Ellis, Sandra Faber, William Fowler, Katherine Freese, George Fuller, Jay Gallagher, Margaret Geller, James Gunn, Alan Guth, Chris Hill, John Huchra, Nick Kaiser, Edward Kolb, David Koo, Masatoshi Koshiba, John Learned, Andre Linde, Grant Mathews, Keith Olive, Jerry Ostriker, Bernard Pagel, Jim Peebles, William Press, Joel Primack, Martin Rees, Hubert Reeves, Bernard Sadoulet, Pierre Sikivie, Joseph Silk, George Smoot, Gary Steigman, Alex Szalay,

Michael Turner, Terry Walker, and David Wilkinson. We also thank Stephen Hawking for writing the Foreword to this book.

We thank the University of Chicago, Fermi National Accelerator Laboratory, and the Stanford Linear Accelerator Center for their support while this book was being written. We also appreciate the hospitality of the Aspen Center for Physics, where a portion of this work was completed.

We are grateful to Roberta Bernstein for her help in the preparation of the manuscript. Jerry Lyons deserves our special thanks for his valuable aid in clarifying our original ideas and revising the manuscript before publication. And we thank Georgia Lee Hadler for her unstinting efforts in converting our often awkward prose and our sketchy drawings into the polished work before you.

This has been a wayward book that took on a life of its own during the more than five years it was being written, rewritten, revised, edited, and published. We are therefore very grateful to the women who lent us support and encouragement at various stages of this project — Sandra Foster, Linda Goodman, and Judy Schramm. We also wish to acknowledge the aid and insight of the late Tobi Saunders of Bantam Books, who was the first person in publishing to recognize the importance of our topic.

Michael Riordan
David N. Schramm

Illustration Credits

page 113 National Optical Astronomy Laboratories

page 121 Lawrence Berkeley Laboratory

page 125 John Mather/NASA

page 137 M. Seldner, B. L. Siebers, E. J. Groth, and P. J. E. Peebles, *Astronomical Journal* 82, 249 (1977)

page 139 Valerie de Lapparent, Margaret Geller, and John Huchra

page 141 Margaret Geller and John Huchra

page 152 Ofer Lahav, Institute of Astronomy, Cambridge UK

page 206 Data from E. J. Groth and P. J. E. Peebles, *Astrophysical Journal* 217, 385 (1977)

page 213 D. Bennett and F. Bouchet

page 231 NASA

page 233 European Southern Observatory

page 237 Cosmic Ray Physics Dept., University of Utah

page 241 Joe Stancmpiano, NGS Staff, and Karl Luttrell, University of Michigan © 1988 National Geographic Society

page 252 Fermi National Accelerator Laboratory

Index